# MONOGRAPHS ON STATISTICS AND APPLIED PROBABILITY

General Editors

**D.R. Cox, V. Isham, N. Keiding, N. Reid, H. Tong, and T. Louis**

1 Stochastic Population Models in Ecology and Epidemiology *M.S. Barlett* (1960)
2 Queues D.R. *Cox and W.L. Smith* (1961)
3 Monte Carlo Methods *J.M. Hammersley and D.C. Handscomb* (1964)
4 The Statistical Analysis of Series of Events *D.R. Cox and P.A.W. Lewis* (1966)
5 Population Genetics *W.J. Ewens* (1969)
6 Probability, Statistics and Time *M.S. Barlett* (1975)
7 Statistical Inference *S.D. Silvey* (1975)
8 The Analysis of Contingency Tables *B.S. Everitt* (1977)
9 Multivariate Analysis in Behavioural Research *A.E. Maxwell* (1977)
10 Stochastic Abundance Models *S. Engen* (1978)
11 Some Basic Theory for Statistical Inference *E.J.G. Pitman* (1979)
12 Point Processes *D.R. Cox and V. Isham* (1980)
13 Identification of Outliers *D.M. Hawkins* (1980)
14 Optimal Design *S.D. Silvey* (1980)
15 Finite Mixture Distributions *B.S. Everitt and D.J. Hand* (1981)
16 Classification *A.D. Gordon* (1981)
17 Distribution-free Statistical Methods, 2nd edition *J.S. Maritz* (1995)
18 Residuals and Influence in Regression R.D. *Cook and S. Weisberg* (1982)
19 Applications of Queueing Theory, 2nd edition *G.F. Newell* (1982)
20 Risk Theory, 3rd edition *R.E. Beard, T. Pentikäinen and E. Pesonen* (1984)
21 Analysis of Survival Data *D.R. Cox and D. Oakes* (1984)
22 An Introduction to Latent Variable Models *B.S. Everitt* (1984)
23 Bandit Problems *D.A. Berry and B. Fristedt* (1985)
24 Stochastic Modelling and Control *M.H.A. Davis and R. Vinter* (1985)
25 The Statistical Analysis of Composition Data *J. Aitchison* (1986)
26 Density Estimation for Statistics and Data Analysis *B.W. Silverman* (1986)
27 Regression Analysis with Applications *G.B. Wetherill* (1986)
28 Sequential Methods in Statistics, 3rd edition
*G.B. Wetherill and K.D. Glazebrook* (1986)
29 Tensor Methods in Statistics *P. McCullagh* (1987)
30 Transformation and Weighting in Regression
*R.J. Carrol and D. Ruppert* (1988)
31 Asymptotic Techniques of Use in Statistics
*O.E. Bardorff-Nielsen and D.R. Cox* (1989)
32 Analysis of Binary Data, 2nd editic

T0203638

# Classification

## 2nd Edition

**A.D. GORDON**

*Reader in Statistics*
*University of St. Andrews, UK*

CRC Press
Taylor & Francis Group
Boca Raton London New York

CRC Press is an imprint of the
Taylor & Francis Group, an **informa** business

A CHAPMAN & HALL BOOK

CRC Press
Taylor & Francis Group
6000 Broken Sound Parkway NW, Suite 300
Boca Raton, FL 33487-2742

First issued in paperback 2019

ISBN-13: 978-0-367-39966-5

**Library of Congress Cataloging-in-Publication Data**

Gordon, A. D.
    Classification / A. D. Gordon. -- 2nd ed.
       p.  cm. -- (Monographs on statistics and applied probability ; 82)
    Includes bibliographical references and index.

    1. Discriminant analysis. 2. Cluster analysis. I. Title.
II. Series.
QA278.65.G67 1999
519.5'35—dc21
                                            99-25287
                                          CIP

Library of Congress Card Number 99-25287

Visit the Taylor & Francis Web site at
http://www.taylorandfrancis.com

and the CRC Press Web site at
http://www.crcpress.com

# Contents

# Preface

Large multivariate data sets can prove difficult to comprehend, and it is useful to have methods of summarizing and extracting information from them. This is becoming increasingly important due to the large amounts of data that are now being collected and stored electronically. The subject of 'classification' is concerned with the investigation of sets of 'objects' in order to establish if they can validly be summarized in terms of a small number of classes of similar objects. The methodology has been used in many different disciplines, and the parts of the subject that are relevant for a particular investigation depend on the nature of the data and the questions of interest: for example, very large data sets will usually be subjected to less elaborate analyses than smaller data sets.

There have been many developments since the publication of the first edition of the book (Gordon, 1981), and the current edition contains a substantial amount of new material. The book has also lost its subtitle, 'Methods for the exploratory analysis of multivariate data', reflecting the fact that cluster validation has been an active area of research in recent years and classification is no longer perceived as being concerned solely with exploratory analyses.

The book is intended not only for statistics undergraduates and postgraduates, but also for research workers in other disciplines who would find the methodology of classification relevant for the analysis of their data. The mathematical level of the presentation is fairly low, although Chapter 6 requires some knowledge of matrix theory, such as may be found in most introductory texts.

Material in the book has been used in a lecture course that has been presented to Honours students in the University of St Andrews for a number of years. Latterly, the course has comprised a mixture of lectures, directed reading, and practical sessions involving the analysis of data using Clustan software.

St Andrews, December 1998                          A. D. Gordon

# Preface

Large multivariate data sets can prove difficult to comprehend, and it is useful to have methods of summarizing and extracting information from them. This is becoming increasingly important due to the large amounts of data that are now being collected and stored electronically. The subject of 'classification' is concerned with the investigation of sets of 'objects' in order to establish if they can validly be summarized in terms of a small number of classes of similar objects. The methodology has been used in many different disciplines, and the parts of the subject that are relevant for a particular investigation depend on the nature of the data and the questions of interest: for example, very large data sets will usually be subjected to less elaborate analyses than smaller data sets.

There have been many developments since the publication of the first edition of the book (Gordon, 1981), and the current edition contains a substantial amount of new material. The book has also lost its subtitle, 'Methods for the exploratory analysis of multivariate data', reflecting the fact that cluster validation has been an active area of research in recent years and classification is no longer perceived as being concerned solely with exploratory analyses.

The book is intended not only for statistics undergraduates and postgraduates, but also for research workers in other disciplines who would find the methodology of classification relevant for the analysis of their data. The mathematical level of the presentation is fairly low, although Chapter 6 requires some knowledge of matrix theory, such as may be found in most introductory texts.

Material in the book has been used in a lecture course that has been presented to Honours students in the University of St Andrews for a number of years. Latterly, the course has comprised a

mixture of lectures, directed reading, and practical sessions involving the analysis of data using Clustan software.

St Andrews, December 1998                        A. D. Gordon

CHAPTER 1

# Introduction

## 1.1 Classification, assignment and dissection

The subject of classification is concerned with the investigation of the relationships within a set of 'objects' in order to establish whether or not the data can validly be summarized by a small number of classes (or clusters) of similar objects. The following examples are intended to illustrate the range of different problems for which the methodology is relevant.

Example 1
Archaeologists have an interest in detecting similarities amongst artifacts, such as ornaments or stone implements found during excavations, as this would allow them to investigate the spatial distribution of artifact 'types' (Hodson, Sneath and Doran, 1966).

Example 2
Plant ecologists collect information about the species of plant that are present in a set of quadrats, listing the species present in each quadrat and possibly also recording a measure of the abundance of each species in the quadrat. One of their aims is to arrange the quadrats in classes such that the members of each class possess some properties which distinguish them from members of other classes (Greig-Smith, 1964).

Example 3
Taxonomists are concerned with constructing classifications which summarize the relationships between taxonomic units of various kinds. The units are usually included in classes which are mutually exclusive and hierarchically nested, as illustrated by the Linnean system (Sneath and Sokal, 1973).

Example 4
Social network analysts investigate the interactions within a set of

individuals, and are interested in identifying individuals who have similar aims or attributes (Arabie and Carroll, 1989).

Example 5

Those who enjoy sampling malt whiskies might be interested in seeing a classification of the distilleries producing these whiskies, as this would enable them to obtain an indication of the range of tastes available by selecting a small number of representatives from each class; using the classification, the whisky producers could identify their commercial competitors (Lapointe and Legendre, 1994).

In these examples, an 'object' corresponds, respectively, to an archaeological artifact, a vegetational quadrat, a taxonomic unit, an individual and a whisky distillery. Objects are generally described by a set of variables: in the first three examples, a variable corresponds, respectively, to a physical property of an artifact, a species of plant and a taxonomic character. However, the distinction between 'object' and 'variable' is not clear-cut: in Example 2, an investigator might be more interested in the relationships between different species of plant, as revealed by their co-occurrences in a set of quadrats. The terminology used in the book refers to the classification of a set of objects, but occasionally such 'objects' might be more accurately described as 'variables'. Some of the methodology in Chapter 6 aims to provide simultaneous graphical representations of the relationships within and between sets of objects and sets of variables.

On occasion, it is not straightforward to define or measure relevant variables. In Example 5, the whiskies analysed by Lapointe and Legendre (1994) were described in terms of several of their characteristics, but in other circumstances one might have to resort to tasting each whisky and providing an overall measure of the perceived difference between each pair of them.

The end result of a classification study is often a *partition* of the set of objects into a set of disjoint classes such that objects in the same class are similar to one another: thus, in Example 1, a class would contain a set of similar archaeological artifacts. However, it might be appropriate to obtain other representations of the objects. Thus, in Example 3, a taxonomic unit could belong successively to a species, a genus, a family, an order, etc., and it would be relevant to obtain a *hierarchical classification*, indicating the relationships between different classes of objects. In Example 4, it could be relevant to summarize the relationships between individuals in an *overlapping classification*, in which different classes could have some but not necessarily all of their members in common. On occasion, it could be informative to provide for each object a set of values which indicate the extent to which it is perceived as belonging to each class.

In classification, in the sense in which the word is used in this book, the classes are not known at the start of the investigation: the number of classes, their defining characteristics and their constituent objects all require to be determined. The word 'classification' has also been used in a different sense, to refer to the assignment of objects to one of a set of already-defined classes. Thus, in pattern recognition or discriminant analysis (Krzanowski and Marriott, 1995; Ripley, 1996) each object is assumed to belong to one of a known number of classes, whose characteristics have been determined using a 'training set', and the aim is to identify the class to which the object should be assigned. In this book, such methodology is referred to as *assignment* or *supervised pattern recognition*, and the word 'classification' is reserved for investigations in which the aim is to determine the classes. Classification is also referred to as unsupervised pattern recognition or unsupervised learning, the methodology often being used in the analysis of training sets as a preliminary to the assignment of new objects. This book has little to say about assignment methodology, although brief comments on it are made in Sections 5.4.2 and 7.3, and a hybrid classification/assignment procedure is described in Section 3.4.2. In recent years classification and assignment procedures have both found many applications in the extraction of information from large data sets, an activity referred to as *data mining* and *knowledge discovery in databases* (Fayyad et al., 1996).

As stated earlier, classification is concerned with seeking valid summaries of data comprising classes of similar objects. An additional requirement for a partition is that the classes be well-separated, i.e. that objects be not only similar to other objects in the same class, but also markedly different from objects in other classes. These desiderata of internal cohesion and external isolation (Cormack, 1971) for a class are illustrated in Fig. 1.1, in which each '+' represents an object described by two quantitative variables. The ideal types of class resemble those depicted in Fig. 1.1(a), but it is clear from the other parts of the Figure that classes need be

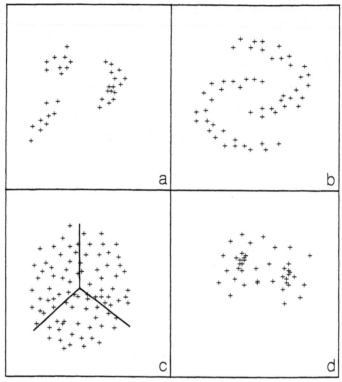

Figure 1.1 *Illustration of the concepts of the cohesion and isolation of classes: (a) classes are cohesive and isolated; (b) classes are isolated but not cohesive; (c) classes are cohesive but not isolated; (d) regions of relatively high density of points surrounded by regions of lower density.*

neither cohesive nor isolated. Fig. 1.1(c) illustrates a *dissection* of the set of objects (Kendall, 1966), since it would be accurate to describe these objects as belonging to a single class; nevertheless it can occasionally be useful to subdivide such classes, and classification algorithms have been used for this purpose. It is more common for data to resemble those depicted in Fig. 1.1(d), in which the regions of higher density might correspond to the nuclei of classes. Given that the objects will generally be described by substantially more than two variables, it should be clear that the accurate detection of the class structure present within a set of objects is not a trivial task.

## 1.2 Aims of classification

Some historical information about the development of theories of classification is presented by Cain (1962), Reyment, Blackith and Campbell (1984) and Sutcliffe (1994). Given the wide range of different applications, one would not expect classification studies always to be undertaken for the same reasons, but two main aims of classification can be identified: data simplification and prediction.

The widespread use of automated methods of data collection and the ease with which large quantities of data can be stored have led to the creation of data sets which are so voluminous and complex that humans carrying out simple inspections of them would find it difficult to extract much information. Large quantities of unassimilated data do not assist understanding: for example, Frawley, Piatetsky-Shapiro and Matheus (1992) report the comment of a manager that 'computers have promised us a fountain of wisdom but delivered a flood of data.' The methodology of classification enables such data sets to be summarized and can help detect important relationships and structure within the data set. If distinct classes of objects are found to exist, they can be named and their properties summarized, thus allowing more efficient organization and retrieval of information and facilitating the assignment of new objects.

Occasionally, the objects being studied comprise the entire set of objects that are of interest and the only concern is with obtaining valid summaries of a single data set. More frequently, summaries of a data set are expected to be relevant for describing a larger collection of objects. Such summaries of data can allow investigators to make predictions or discover hypotheses to account for the structure in the data. These predictions can be at various levels of sophistication. At the simplest level, one could predict properties which have not been recorded for an object, by comparing it with other objects which it is found to resemble: in Example 1, the missing features of a damaged artifact could be predicted in this manner. Similarly, one could predict that objects placed in the same class might resemble one another in variables which were not used to obtain the classification: thus, Watson, Williams and Lance (1966) state that 'those who need a new source of an unusual plant product turn first to the taxonomic relatives of species known to produce it.'

More generally, investigators can predict that new objects 'collected' in similar circumstances are likely to possess similar properties to those already studied and can then take decisions or plan actions on the basis of such predictions: for example, past behaviour of a business's clients can be used to suggest how the business should plan for the future.

At a deeper level, the results of classification studies could enable investigators to formulate general hypotheses to account for the observed data. Probably the most famous examples of such classifications are the Linnean system in biology and the periodic table of chemical elements – both of which were constructed without recourse to modern automated methods of classification!

In the above discussion, a classification provides an assertion about the data under investigation. Given the ability of the human mind to think up *post hoc* justifications of conclusions, even those which should properly be regarded as dubious, it is clearly important that the results of a classification be subject to a process of validation; a fuller discussion of this topic is presented in Chapter 7.

It should be noted that there are many different ways in which a set of objects could be classified. Thus, one could classify a book by its contents, its author or its size, and though this last criterion may have little to commend it, it should at least separate many atlases and art books from other types of book (Good, 1965). Similarly, an investigator might be more interested in classifying trees, shrubs and herbs by their medicinal properties than by more commonly used taxonomic criteria (Gilmour, 1937). Hence, any classification that is obtained should be related to features in which the investigator is interested and it is quite possible for there to be several different classifications of the same set of objects. The implication is that the investigator should give careful thought to selecting the set of variables that will be used to describe the objects; further comments on this topic are postponed to Chapter 2.

## 1.3 Stages in a numerical classification

This section provides an overview of the material in the book. A simplified sketch of stages in a classification study is given in Fig. 1.2, although it should be noted that some studies make use of more specialized methodology or are concerned with only a subset of the stages shown in the Figure. Many decisions require to be made before the successful conclusion of a classification study and

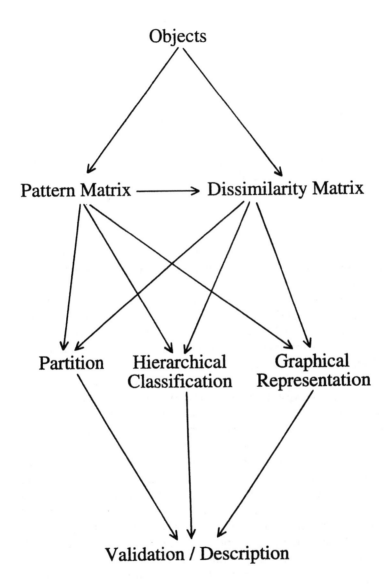

Figure 1.2 *Stages in the classification of objects.*

Table 1.1 *Questions that require to be answered in the course of a classification study and sections of the book that address these questions.*

| | | |
|---|---|---|
| 1. | How should the objects for analysis be selected? | 1.3, 3.4.2 |
| 2. | Which variables should be used to describe the objects? | 1.2, 2.3, 2.4 |
| 3. | Should any standardization or differential weighting of variables be undertaken? | 2.3, 2.4 |
| 4. | How should a relevant measure of dissimilarity be constructed from a pattern matrix? | 2.4 |
| 5. | Which clustering and graphical procedures should be used in the analysis of data? | 3.6, 4.3, 6.3 |
| 6. | How should the results of the study be validated and summarized? | 7.1 – 7.3 |

insufficient attention has been paid to what can be termed strategic issues: much has been written on how investigations *can* be conducted; there has been less advice on how investigations *should* be conducted. A list of questions that need to be answered is given in Table 1.1, together with sections of the book where these questions are addressed; again, it is stressed that not all questions are relevant in every investigation. Further discussion of this topic is given in Milligan (1996).

The first question concerns the specification of the set of objects on which the investigation is to be carried out. One should consider whether interest resides solely in the set of objects to be classified or whether these objects are to be regarded as a sample from a larger collection (or 'population') of objects, the properties of the population being what are of prime importance. Thus, in Example 5 the only data that are of interest are those describing the malt whiskies produced by (just over 100) Scottish distilleries, whereas in Example 2 a classification of a set of quadrats should provide more general information about the vegetation in a region.

If information about the population of interest is to be provided by a sample of objects, the question arises as to how that sample should be selected. It is difficult to make specific recommendations that will be relevant in all subject areas. As a general observation, the aim should be to obtain a set of objects that covers the entire range of variability encompassed by the population. If the population contains classes of similar objects, one would like the sample to include a reasonable number of representatives from each of these classes. This implies that simple random sampling, even if possible, might not be appropriate: it could be more helpful to obtain a higher proportional contribution from smaller classes in the population, so as to obtain a more reliable description of them, i.e. it is usually considered more important to obtain information about the classes rather than estimates of their relative frequencies. To assist in the description and interpretation of classes, it can occasionally be relevant to augment the data set by inserting several 'ideal' objects and noting how they fit into the classification.

A further question concerns whether or not it is necessary to collect information about objects other than those contained in the set that will be classified (and possibly a set that can be used to validate the results). A crucial distinction appears to be whether or not the main characteristics of an object can be known or accurately guessed without detailed investigation. Sometimes there may be a proxy variable that reflects the range of variation of objects in the population, such as geographical location in some remote sensing studies. Occasionally, experienced investigators can rapidly assess (potential) objects in the population in order to select a sample (e.g. in Example 2), but there are clearly dangers of obtaining biased information by using such an approach. However, particularly with the increased use of automated methods of data collection and storage, it is often the case that complete information is available on a large set of objects and it is relevant to consider selecting a subset of these objects for classification. Ways of extracting relevant subsets are described in Section 3.4.2.

The set of $n$ (say) objects for which a classification is sought is usually described in one of two formats: a *pattern* (or *profile*) *matrix* or a *(dis)similarity matrix*. A pattern matrix is an $n \times p$ objects × variables matrix, whose $(i, k)$th element provides a value or category for the $k$th variable describing the $i$th object ($i = 1, ..., n; k = 1, ..., p$). A dissimilarity matrix is a symmetric $n \times n$ matrix whose $(i, j)$th element provides a measure of the dissimilarity between the $i$th and $j$th objects ($i, j = 1, ..., n$); alternatively, an $n \times n$ similarity matrix holds measures of the similarities between each pair of objects. A widely-adopted terminology (Tucker, 1964) distinguishes between *modes* and *ways* of data matrices: the modes are the distinct types of entity (e.g. objects, variables) and the ways refer to the subscripts of entries in the matrix (e.g. a matrix with only rows and columns is two-way). Thus, a pattern matrix contains two-way two-mode data and a dissimilarity matrix contains two-way one-mode data.

As indicated in Fig. 1.2, dissimilarity matrices can be specified directly or be derived from pattern matrices. Chapter 2 discusses the construction of dissimilarity matrices, providing answers to questions 2 – 4 in Table 1.1. The input to most clustering algorithms is a pattern matrix or a (dis)similarity matrix. The most common forms of output are a partition of the set of objects and a hierarchical classification, and these are described in Chapters 3 and 4, respectively. These chapters contain comments on properties of the output obtained from various classification algorithms, providing some answers to the fifth question in Table 1.1. However, partitions and hierarchical classifications are not the only ways in which it can be informative to summarize data: Chapter 5 presents a miscellaneous collection of methodology which is relevant for providing other types of classification or for analysing more general kinds of data.

It can be misleading to impose a classification on a set of objects. A valuable adjunct to a classification is a graphical representation in which each object is represented by a point, objects that are similar to one another being represented by points that are close together. Fig. 1.1 provides such representations for two-dimensional data sets. If the pattern matrix were of a higher dimensionality or the data were provided in a dissimilarity matrix, it would still be useful to obtain low-dimensional configurations of points that provided good approximations of the relationships between the objects, and then assess whether or not any class structure were apparent in such representations. Procedures for obtaining such low-dimensional configurations of points and a discussion of their properties are presented in Chapter 6.

The final stages in Fig. 1.2 concern the validation of any class structure that has been indicated as present and the description of defining characteristics or relevant properties of classes, providing answers to the sixth question in Table 1.1. These topics, which have received insufficient attention in many classification studies, are addressed in Chapter 7.

## 1.4 Data sets

The book includes analyses of data using the methodology that is described. Some small data sets are described in the text and larger data sets are available by anonymous ftp from **pi.dcs.st-and.ac.uk** in directory **pub/adg**. Also available there are comments on software used to carry out the analyses and some more general comments on available software, which it is intended to update at regular intervals.

The remainder of this section provides some information about the larger data sets.

### 1.4.1 Abernethy Forest data

As an aid to identifying past vegetation, pollen analysts take vertical cores of sediment, usually from lake beds or peat bogs, and identify the pollen composition in the sediment at various positions down the core. Two such data sets were taken from Abernethy Forest in north-east Scotland. The first (Birks, 1970) comprises 41 samples and the second (Birks and Mathewes, 1978) comprises 49 samples; in both cases, samples are numbered from youngest to oldest down the core. Many of the pollen types are present in very small quantities, and attention is restricted to the nine quantitatively most important pollen types. Each of the 90 samples is described by the proportions of its (reduced) pollen sum that belong to each of these nine pollen types. The second data set is referred to as the Abernethy Forest 1974 data, as it was collected in that year. Analyses of these data are presented in Sections 3.2 – 3.5, 4.2, 5.2, 6.2, 6.4, 6.5, 7.2 and 7.3.

### 1.4.2 Acoustic confusion data

Miller and Nicely (1955) collected data on the acoustic confusion between sixteen English consonants. Subjects listened to sixteen syllables formed by adding 'a' after the consonant, recording the syllable which they believed had been spoken. The sixteen consonants are given by the beginning of the following sounds: *pa, ka, ta, fa, thin, sa, sha, ba, va, that, da, ga, za, zha, ma, na.* Shepard (1972) aggregated the data collected under six different experimental conditions and symmetrized them, to obtain a matrix containing a measure of the similarity between each pair of consonants. Analyses of these data are presented in Sections 4.2, 4.4, 5.3, 6.2 – 6.4, 7.2 and 7.3.

### 1.4.3 Diday and Govaert's data

Diday and Govaert (1977) generated 50 observations from each of three bivariate normal distributions, whose mean vectors and covariance matrices were:

$$\mu_1 = \begin{pmatrix} 0 \\ 0 \end{pmatrix}, \ \mu_2 = \begin{pmatrix} 0 \\ 3 \end{pmatrix}, \ \mu_3 = \begin{pmatrix} 4 \\ 3 \end{pmatrix}$$

$$\Sigma_1 = \begin{pmatrix} 4 & 1.7 \\ 1.7 & 1 \end{pmatrix}, \Sigma_2 = \begin{pmatrix} 0.25 & 0 \\ 0 & 0.25 \end{pmatrix}, \Sigma_3 = \begin{pmatrix} 4 & -1.7 \\ -1.7 & 1 \end{pmatrix}$$

Analyses of these data are presented in Sections 3.2 and 5.1.

### 1.4.4 European fern data

Birks (1976) collected information about the spatial distribution of 144 species of fern in Europe. Sixty-five geographical regions were described by the presence or absence of each species of fern. Sixty-one of these regions comprise an area of 6 degrees longitude by 5 degrees latitude. Each of these areas has been coded by three digits, the first of which indicates the 'northness' of the region and the last two of which indicate its 'eastness'; region 101 covers the area of Portugal and Spain lying between 40 degrees and 45 degrees north and between 12 degrees and 6 degrees west. The region coded 134 is the union of regions 103 and 104, and the other three regions comprise the islands of Iceland (coded **ice**), Svalbard (**sva**) and the Azores (**azo**). Analyses of these data are presented in Sections 4.2 and 5.4.

### 1.4.5 Flint arrowheads data

J. B. Kenworthy recorded the physical characteristics of a collection of leaf-shaped flint arrowheads recovered during archaeological investigations in north-east Scotland. A subset of 14 of these arrowheads is used in Section 2.4 to illustrate the construction of relevant measures of dissimilarity. Each arrowhead is described by five variables that record aspects of its size and shape. A matrix of the perceived dissimilarities between each pair of arrowheads was also provided by an assessor.

## 1.4.6 Kinship terms data

Rosenberg and Kim (1975) collected data on fifteen kinship terms (grandfather, grandmother, grandson, granddaughter, brother, sister, father, mother, son, daughter, nephew, niece, uncle, aunt, cousin). A subset of their data comprises partitions provided by 85 female undergraduates at Rutgers University, who were asked to sort the kinship terms into categories 'on the basis of some aspect of meaning.' The data are given in Rosenberg (1982, Table 7.1), with the 'nephew' and 'niece' columns transposed. Analyses of these data are presented in Sections 4.5 and 5.6.

# CHAPTER 2

# Measures of similarity and dissimilarity

## 2.1 Introduction

Many of the methods of analysis described in the book assume that the relationships within a set of $n$ objects are described in an $n \times n$ matrix containing a measure of the similarity $s_{ij}$ or dissimilarity $d_{ij}$ between the $i$th and $j$th objects for each pair of objects $(i, j)$ $(i, j = 1, ..., n)$. Three-way measures of the (dis)similarity of triples of objects and methodology relevant for the analysis of such data have also been proposed (Hayashi, 1972; Joly and Le Calvé, 1995; Daws, 1996), but the book concentrates on the analysis of pairwise (dis)similarities. Most of the presentation is in terms of dissimilarities. Measures of similarity can often be transformed into measures of dissimilarity (e.g. $d_{ij} = k - s_{ij}$ for some constant $k$), but in some problems there is not a straightforward transformation linking perceived similarity and dissimilarity (Tversky, 1977). Self-similarities $s_{ii}$ $(i = 1, ..., n)$ are often assumed to take a common value, but this need not be the case.

Attention is restricted to measures of pairwise dissimilarity $d_{ij}$ that satisfy the following three conditions for all $i, j = 1, ..., n$:

$$(i)\ d_{ij} \geq 0; \quad (ii)\ d_{ii} = 0; \quad (iii)\ d_{ij} = d_{ji}.$$

The third condition rules out asymmetric measures of dissimilarity and, taken in conjunction with condition (ii), implies that a matrix of dissimilarities is fully specified by providing the $n(n-1)/2$ values in its lower triangle. Asymmetric difference matrices $(\Delta_{ij})$ do occur: for example, in the analysis of confusion data on Morse code symbols, $\Delta_{ij}$ could denote the complement of the proportion of times that the $i$th symbol was sent and the receiver recorded the $j$th symbol. Information about the perceived resemblances between Morse code symbols could be obtained from a classification based on a matrix of dissimilarities $(d_{ij})$ defined by $d_{ij} = (\Delta_{ij} + \Delta_{ji})/2$

(Shepard, 1963), but such an analysis might miss important features of the data. Methods of analysing asymmetric difference data have been proposed (e.g. Hubert, 1973a; Gower, 1977; Brossier, 1982; Ferligoj and Batagelj, 1983; Kiers and Takane, 1994), but the methodology described in this book for the analysis of pairwise dissimilarity data assumes that the dissimilarities are symmetric.

Some methods of analysis assume, either explicitly or implicitly, that the objects can be represented by points in a Euclidean space and it is relevant to investigate if dissimilarity measures are compatible, or can be made compatible, with this assumption.

A dissimilarity matrix $\mathbf{D} = (d_{ij})$ is said to be *metric* if it satisfies the triangle inequality:

$$d_{ij} \leq d_{ik} + d_{kj}$$

for all triples of objects $(i, j, k)$.

$\mathbf{D}$ is said to be *Euclidean* if there exists a configuration of points in Euclidean space $\{P_i \ (i = 1, ..., n)\}$ with the distance between $P_i$ and $P_j$ equal to $d_{ij}$.

If $\mathbf{D}$ is Euclidean it is also metric, but the converse does not hold: a counter example presented by Gower and Legendre (1986) is based on modifying a configuration of four points in which points $P_1, P_2$ and $P_3$ are located at the vertices of an equilateral triangle of side 2, and $P_4$ is located at the centre of the triangle. This configuration has interpoint distances $d_{ij} = 2 \ (i, j \in \{1, 2, 3\}, i \neq j)$ and $d_{4j} = 2/(\sqrt{3}) \approx 1.15 \ (j = 1, 2, 3)$. If the distances between the first three points are unaltered, but $d_{4j}$ is modified to take any value in the interval $(1, 2/(\sqrt{3}))$, $\mathbf{D}$ is metric but not Euclidean.

Gower (1966) shows that if $\mathbf{S} = (s_{ij})$ is a non-negative definite similarity matrix for which $s_{ii} = 1 \ (i = 1, ..., n)$, the dissimilarity matrix whose elements are $d_{ij} \equiv (1 - s_{ij})^{1/2}$ is Euclidean. A dissimilarity matrix can also be transformed to be Euclidean: for any dissimilarity matrix $\mathbf{D} = (d_{ij})$ there exist constants $c_1$ and $c_2$ such that the matrix with elements $(d_{ij}^2 + c_1)^{1/2}$ is Euclidean (Lingoes, 1971) and the matrix with elements $(d_{ij} + c_2)$ is Euclidean (Cailliez, 1983). However, when some of the values in a metric (or Euclidean) dissimilarity matrix are missing and are imputed by some procedure, the resulting dissimilarity matrix need not be metric (or Euclidean). Further investigations of metric and Euclidean properties of dissimilarities are reported by Fichet and Le

Calvé (1984), Gower and Legendre (1986) and Cailliez and Kuntz (1996).

As indicated in Fig. 1.2, the set of objects can be described by a pattern matrix. If a relevant classification algorithm requires the data to be presented as a matrix of (dis)similarities, it is necessary to transform the pattern matrix to meet this requirement. This chapter discusses problems in specifying the variables to be used in describing objects and describes ways in which matrices of pairwise (dis)similarities can be derived from pattern matrices, addressing difficulties that can arise during this exercise and suggesting how to construct a measure of dissimilarity that is appropriate for the set of objects under investigation.

## 2.2 Selected measures of similarity and dissimilarity

Many different measures of pairwise similarity and dissimilarity have been proposed in the classification literature. Some of these are of dubious or limited usefulness and others are closely related to one another. A selection of measures is presented in this section; later sections in the chapter offer some advice to investigators about how to select measures that are relevant for their data.

The terminology in this section concentrates on describing measures of the (dis)similarity between a pair of objects described by a set of variables, but measures relevant in the situation in which each 'object' is a group of objects are also presented; a discussion of other measures of dissimilarity relevant for symbolic data is postponed to Section 5.4. Much of the following material is also relevant for the construction of measures of the (dis)similarity between pairs of variables; a more detailed discussion of this latter type of comparison is given by Anderberg (1973, Chapter 4). Most of the measures presented are particularly relevant for comparing objects that are described by a single type of variable and the measures are categorized in this way in subsections 2.2.1 – 2.2.3.

### 2.2.1 Binary variables

A binary variable can belong to one of only two states, denoted by '+' and '−' say. Consider a set of objects described by $p$ binary variables, e.g. presence (+) or absence (−) of each of $p$ species of plants in an ecological quadrat. For the pair of objects $(i, j)$, let $a$ denote the number of variables that are + for both objects, $b$

denote the number of variables that are + for the $i$th object and − for the $j$th object, $c$ denote the number that are − for the $i$th object and + for the $j$th object, and $d$ denote the number that are − for both objects; $p = a + b + c + d$. There are many measures of the (dis)similarity between the $i$th and $j$th objects which are functions of $a$, $b$, $c$ and $d$. Two of the more common ones are:

M1: Simple matching coefficient

$$s_{ij} \equiv (a + d)/p; \quad d_{ij} \equiv (b + c)/p.$$

M2: Jaccard's coefficient

$$s_{ij} \equiv a/(a + b + c); \quad d_{ij} \equiv (b + c)/(a + b + c),$$

with $s_{ij} \equiv 1$ and $d_{ij} \equiv 0$ if $a = b = c = 0$.

For both of these coefficients, the matrix $(s_{ij})$ is non-negative definite (Gower and Legendre, 1986) and hence the matrix with elements $(d_{ij}^{1/2})$ is Euclidean.

Other coefficients which are functions of $a$, $b$, $c$ and $d$ are given by Hubálek (1982), Gower and Legendre (1986) and Baulieu (1989, 1997). Some of these coefficients are monotonically related to (i.e. provide the same ordering of the $n(n-1)/2$ (dis)similarities as) M1 or M2. Baulieu (1989, 1997) presents sets of axioms one might require of any dissimilarity coefficient for objects described by binary variables and proves that these axioms uniquely define families of coefficients, for example

$$(b + c)/(\alpha a + b + c + \delta d), \text{ where } \alpha, \delta > 0$$

and

$$(b + c)/(\alpha a + b + c), \text{ where } \alpha > 0.$$

### 2.2.2 Multistate nominal and ordinal variables

Multistate variables are categorical variables for which there are more than two states or categories. If the states are (resp., are not) ordered, the variable is termed ordinal (resp., nominal). In considering measures of dissimilarity based on nominal or ordinal variables, attention is focussed on specifying the contribution made to such a measure by a single variable. Overall measures of the dissimilarity between a pair of objects are obtained by summing such contributions over all the variables.

Disagreement indices can be defined between each pair of states of the categorical variable: let $\delta_{klm}$ $(\geq 0)$ denote the disagreement between the $l$th and $m$th state of the $k$th variable ($k = 1, ..., p; 1 \leq l, m \leq c_k$, where $c_k$ denotes the number of states); clearly, $\delta_{klm} = \delta_{kml}$. The indices for nominal variables would normally be defined by $\delta_{klm} = 1$ if $l \neq m$ and $\delta_{kll} = 0$ ($l, m = 1, ..., c_k$), although more general formulations are possible if the investigator has reason to believe that these would be more appropriate.

For an ordinal variable, for which it is assumed that the states are numbered in the correct order, the set $(\delta_{klm})$ should satisfy monotonicity conditions:

$$\delta_{klm} < \delta_{klr} \text{ if } l > m > r \text{ or } l < m < r \ (k = 1, ..., p).$$

It is convenient to assume that the largest disagreement, $\delta_{k1c_k}$, takes the value 1 ($k = 1, ..., p$). In the absence of more detailed information, $\delta_{klm}$ could be defined to be proportional to $| l - m |^r$, with $r = 1$ a common exponent. If the states of the ordinal variable have been obtained by sectioning a quantitative variable, more elaborate scoring methods might be appropriate if information is available about the distribution of the underlying quantitative variable.

The contribution to the dissimilarity $d_{ij}$ between the $i$th and $j$th objects that is made by the $k$th variable is defined by $d_{ijk} = \delta_{klm}$ if the $k$th variable is in state $l$ for the $i$th object and state $m$ for the $j$th object; the contribution to the similarity $s_{ij}$ can be defined by $s_{ijk} = 1 - d_{ijk}$. The overall measure of (dis)similarity is defined by:

M3: General nominal/ordinal (dis)similarity

$$s_{ij} \equiv \Sigma_{k=1}^{p} s_{ijk}; \ \ d_{ij} \equiv \Sigma_{k=1}^{p} d_{ijk}.$$

These measures could be divided by the number of variables, $p$, to ensure that they lie between 0 and 1.

### 2.2.3 Quantitative variables

Let $x_{ik}$ denote the value that the $k$th quantitative variable takes for the $i$th object ($i = 1, ..., n; k = 1, ..., p$). A family of dissimilarity measures, indexed by a parameter $\lambda$, is available:

M4: Minkowski metrics

$$d_{ij} \equiv (\Sigma_{k=1}^{p} w_k^{\lambda} \mid x_{ik} - x_{jk} \mid^{\lambda})^{1/\lambda} \ (\lambda \geq 1),$$

where $\{w_k \ (k = 1, ..., p)\}$ are non-negative weights associated with the variables, allowing standardization and weighting of the original variables, as discussed later in the chapter. The most commonly used of these metrics have $\lambda = 1$ or 2.

M5: City block metric

$$d_{ij} \equiv \Sigma_{k=1}^{p} w_k \mid x_{ik} - x_{jk} \mid .$$

M6: Euclidean distance

$$d_{ij} \equiv (\Sigma_{k=1}^{p} w_k^2 (x_{ik} - x_{jk})^2)^{1/2} .$$

Measures M4 – M6 are often standardized, e.g. to ensure that $d_{ij}$ is bounded above by 1 one can define $w_k = (pR_k)^{-1}$, where $R_k$ denotes the range of values taken by the $k$th variable. A dissimilarity measure with an inbuilt standardization is:

M7: Canberra metric (Lance and Williams, 1966a)

$$d_{ij} \equiv \Sigma_{k=1}^{p} \mid x_{ik} - x_{jk} \mid /(\mid x_{ik} \mid + \mid x_{jk} \mid),$$

where the summand is defined to be zero if $x_{ik} = 0 = x_{jk}$. This measure is very sensitive to small changes close to $x_{ik} = 0 = x_{jk}$ and can be less reliable if the $(x_{ik})$ are sample estimates of some quantities; however, this sensitivity near zero values makes it an appropriate generalization of dissimilarity measures based on binary presence/absence variables. The measure has also been divided by $p$, to ensure that $d_{ij}$ lies between 0 and 1.

Sometimes the precise values taken by the variables describing an object are important only to the extent that they provide information about the relative magnitudes of the different variables. Thus, the values of the variables describing an object define a vector with $p$ components and interest is restricted to the comparison of the directions of the vectors. Two correlation-type measures of similarity are given below; they both measure the cosine of the angle between two vectors, with the vectors being measured, respectively, from the origin and from the 'mean' of the data.

M8: Angular separation

$$s_{ij} \equiv \frac{\Sigma_{k=1}^{p} x_{ik} x_{jk}}{(\Sigma_{k=1}^{p} x_{ik}^2 \Sigma_{l=1}^{p} x_{jl}^2)^{1/2}}.$$

M9: Correlation coefficient

$$s_{ij} \equiv \frac{\Sigma_{k=1}^{p} (x_{ik} - \bar{x}_{i.})(x_{jk} - \bar{x}_{j.})}{(\Sigma_{k=1}^{p} (x_{ik} - \bar{x}_{i.})^2 \Sigma_{l=1}^{p} (x_{jl} - \bar{x}_{j.})^2)^{1/2}},$$

where

$$\bar{x}_{i.} \equiv \Sigma_{k=1}^{p} x_{ik}/p.$$

Correlations take values between –1 and 1, and these measures can be transformed to take values between 0 and 1 by defining $s_{ij}^* = (1 + s_{ij})/2$.

Further measures of the pairwise (dis)similarity between objects described by quantitative variables are given by Gower (1985) and Gower and Legendre (1986).

### 2.2.4 Mixed variables

Each of the (dis)similarity measures described in Sections 2.2.1 – 2.2.3 is particularly appropriate for objects described by a single type of variable. In some cases, the variables describing a set of objects are of several different types: for example, an archaeological artifact can be described in terms of its weight (quantitative variable), amount of wear (ordinal variable), type of rock (nominal variable) and presence/absence of grip (binary variable). A general measure, which is relevant for variables of all of these types and hence can be used to compare objects described by variables of mixed type, is:

M10: General (dis)similarity coefficient (Gower, 1971a)

$$s_{ij} \equiv \Sigma_{k=1}^{p} w_{ijk} s_{ijk}/\Sigma_{k=1}^{p} w_{ijk}; \quad d_{ij} \equiv \Sigma_{k=1}^{p} w_{ijk} d_{ijk}/\Sigma_{k=1}^{p} w_{ijk},$$

where $s_{ijk}$ (resp., $d_{ijk}$) denotes the contribution to the measure of similarity (resp., dissimilarity) provided by the $k$th variable, and

$w_{ijk}$ is usually 1 or 0 depending on whether or not the comparison is valid for the $k$th variable. The values of $s_{ijk}$ or $d_{ijk}$ can be defined for the different types of variable along the lines indicated earlier; for quantitative variables, Gower (1971a) advocated use of

$$s_{ijk} = 1 - |x_{ik} - x_{jk}|/R_k,$$

where $R_k$ is the range of values taken by the $k$th variable, either in the set of $n$ objects or in some larger population.

### 2.2.5 Comparing groups of objects

Many measures have been proposed of the affinity or divergence of two populations which are described by density functions $f_i$ and $f_j$ (say) with respect to a suitable measure $\nu$. Two examples are:

M11: Bhattacharyya (1943)

$$s_{ij} \equiv \int f_i^{1/2} f_j^{1/2} d\nu$$

M12: Matusita (1956)

$$d_{ij} \equiv [\int (f_i^{1/2} - f_j^{1/2})^2 d\nu]^{1/2}$$

Other measures are described by Gower (1985) and Papaioannou (1985). M12 has also been referred to as Hellinger's distance, its properties being discussed by Le Cam (1970). The measures M11 and M12 are related by

$$d_{ij} = [2(1 - s_{ij})]^{1/2}.$$

When $f_i$ and $f_j$ are multivariate normal density functions with mean vectors $\mu_i$ and $\mu_j$ and common covariance matrix $\Sigma$, M11 and M12 are monotonically related to the Mahalanobis distance,

$$\Delta_{ij}^2 \equiv (\mu_i - \mu_j)'\Sigma^{-1}(\mu_i - \mu_j)$$

(Bhattacharyya, 1943; Matusita, 1967).

These measures of population affinity or divergence can be used to provide measures of the (dis)similarity between pairs of groups of objects by replacing population parameters by estimates of these quantities provided by the groups. For quantitative variables, the relevant Mahalanobis distance is obtained by replacing $\mu_i$ (resp.,

$\mu_j$) by the sample mean vector within the $i$th (resp., $j$th) group and by replacing $\Sigma$ by the pooled sample covariance matrix.

For a set of categorical variables, the $k$th of which has $s_k$ states ($k = 1, ..., p$), a new composite variable with a total of $t \equiv \prod_{k=1}^{p} s_k$ states is specified by combinations of states of the original categorical variables; e.g. the two categorical variables describing trees $V_1$: leaf shape $\in$ {heart-shaped, pointed oval} and $V_2$: bark texture $\in$ {smooth, rough}   give rise to the four composite states {leaf shape = heart-shaped & bark = smooth}, {leaf shape = heart-shaped & bark = rough}, {leaf shape = pointed oval & bark = smooth} and {leaf shape = pointed oval & bark = rough}. Let the probabilities of these composite states in the $i$th population be denoted by $\pi_{ir}$ ($r = 1, ..., t$), where $\pi_{ir}$ is estimated by the relative frequency of occurrence, $\hat{\pi}_{ir}$, of the $r$th composite state in the $i$th group. The square root transformation used in M11 and M12 implies that each group can be represented by a point on the surface of a unit hypersphere (since

$$\Sigma_{r=1}^{t}(\hat{\pi}_{ir}^{1/2})^2 = 1).$$

Bhattacharyya's similarity between the $i$th and $j$th groups of objects is the cosine of the angle between the vectors linking the origin to the $i$th and $j$th points on the surface of the hypersphere, and Matusita's distance is the chord distance between these two points.

If the groups of objects should be regarded as samples from a population or larger collection of objects, the values of the measures of (dis)similarity described in this subsection become more reliable as the sizes of the groups increase.

## 2.3 Some difficulties

### 2.3.1 Selection and standardization of variables

Sometimes, the variables that should be used to describe objects in a classification study can be specified without controversy. In Example 2 in Chapter 1, vegetational quadrats are defined in terms of the plant species contained in them, and while there can be discussion about whether it is appropriate to describe a species in terms of its percentage cover in the quadrat, or some ordinal variable based on the percentage cover, or simply as present in or absent from the quadrat, the specification of the set of variables should be reasonably uncontentious. On other occasions, however, variables have to be selected from a large set of possible variables: in Example 1, the selection of relevant characteristics describing an archaeological artifact may be far from straightforward. This exercise of specifying appropriate variables is referred to in the pattern recognition literature as feature extraction.

In the past, some investigators have taken the view that a large number of variables should be used, in order not to exclude ones that are possibly relevant. While important differences between objects may still emerge from this exercise, there are dangers in such an inclusive strategy. This is illustrated in Fig. 2.1, which provides a plot of a set of 42 objects in terms of their first two variables; the labels of the objects should be disregarded for the present. It is clear from Fig. 2.1 that these data comprise three well-separated classes of objects. However, the objects were also described by six other variables, each of which was generated randomly from a normal distribution with mean 0 and variance approximately equal to the sample variances of the first two variables. This eight-dimensional data set was then partitioned into three classes using the sum of squares clustering criterion described in Section 3.1; the labels (1, 2 or 3) attached to each object in Fig. 2.1 indicate the class membership in the optimal partition into three classes that was found. It is evident that the class structure that was present in the two-dimensional data set has been obscured (but not completely hidden) in the eight-dimensional data set, 12 of the 42 objects having been 'incorrectly' classified. Other examples of the masking of structure by the inclusion of irrelevant variables are presented by Milligan (1980), DeSarbo and Mahajan (1984) and Fowlkes, Gnanadesikan and Kettenring (1988); the latter paper also describes an algorithm for seeking subsets of variables that lead to the determination of a clear cluster structure.

Fig. 2.1 illustrates the importance of giving careful consideration to the choice of variables that are used to describe objects and of including only those that are relevant for the purpose at hand; it is, of course, possible for there to be more than one relevant classification of a set of objects, based on different (but possibly overlapping) sets of variables.

Even if one can be reasonably confident that appropriate variables have been used in the description of a set of objects, it is still necessary to give consideration to the relevance of standardizing and/or differentially weighting them, and often also to the

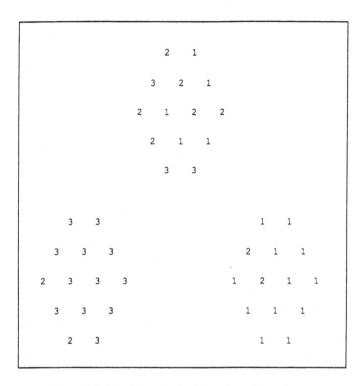

Figure 2.1 *Partition of 42 objects into three classes.*

construction from them of measures of the pairwise (dis)similarity
between objects.

The incentive for carrying out standardization of variables has
often come from the fact that two variables can have markedly
different amounts of variability across a data set. Standardization
has also provided investigators with a solution to the problem of
the relevant units of measurement to use for quantitative variables
(e.g. centimetres, millimetres or inches?). The standardization has
generally been with respect to the data set under investigation:
for example, the values taken by each quantitative variable can be
standardized by dividing them by their standard deviation or by
the range of values they take in the data set. Restricting attention
to quantitative variables, Milligan and Cooper (1988) carried out
a detailed study of the performance of several different methods of

transformation in recovering the correct structure of a large number of simulated data sets, noting that standardization by dividing
by the range outperformed the other methods of standardization
that they investigated. Such simulation studies are valuable, but
their results should be interpreted with care: the fact that a numerical procedure has a success rate that is high relative to its
competitors in the analysis of many data sets does not mean that
it is necessarily the most appropriate method of analysis for a particular data set that is under investigation.

Standardization is also relevant when the variables describing
an object are of different types, and measures of (dis)similarity
like M10 in Subsection 2.2.4 can combine suitably standardized
contributions from several different types of variable.

The question of whether or not to standardize variables is, in
fact, subsumed within the more general problem of the differential
weighting of variables. Further, the selection of variables to include
in an analysis can be regarded as an extreme case of weighting,
with discarded variables being assigned zero weight. Algorithms
that select weights so as to ensure that class structure is as clear-
cut as possible have been proposed by DeSarbo et al. (1984), De
Soete, DeSarbo and Carroll (1985) and De Soete (1986). An alternative approach, that addresses only the selection of weights in
the construction of dissimilarity measures, is described in the next
section.

### 2.3.2 Missing values

Missing values are a common feature of some classification studies: for example, archaeological artifacts might be damaged, pre-
venting the recording of some variables, or questionnaires might
be returned with the answers to some questions not completed.
A detailed discussion of strategies for handling missing values is
presented by Little and Rubin (1987). A reluctance to make many
assumptions about the data set under investigation restricts the
number of approaches applicable in classification studies.

As a general strategy, one can make use of the general (dis)simil-
arity coefficient M10: if the $k$th variable is not recorded for either
or both of the $i$th and $j$th objects, $w_{ijk}$ is set equal to zero and the
overall (dis)similarity between the $i$th and $j$th objects is obtained
by averaging over a smaller number of variables. This approach
assumes that the contribution to the (dis)similarity between two

objects that would have been provided by an incompletely recorded variable is equal to the weighted mean of the contributions provided by the variables for which complete information is available. If an object has many variables for which information is missing, it would be relevant to consider removing the object from the data set because its (dis)similarities with other objects are likely to be unreliable.

It is often relevant to estimate missing values. This requires assumptions about the mechanism that led to a value being missing, e.g. that the probability that a value is missing is not dependent on what that value would have been if it had been observed (Little and Rubin, 1987, Chapter 1).

Suppose that the $i$th object lacks a reading $x_{im}$ for the $m$th variable and let $C_i$ denote a set of objects to be used to provide an estimate $\hat{x}_{im}$ of $x_{im}$. The set $C_i$ could comprise the $r$ nearest neighbours of the $i$th object, or the class to which the $i$th object belongs in the classification of the complete set of objects; in each case, it is assumed that the $m$th variable has been recorded for all of the objects in $C_i$. If the $m$th variable is a nominal variable, $\hat{x}_{im}$ can be defined to be the state which occurs most frequently in the set $C_i$. For a missing ordinal variable, one could specify the state corresponding to the median observation within $C_i$. For a quantitative variable, $\hat{x}_{im}$ can be defined to be the mean or median of the values in $C_i$.

These methods of imputation do not make any direct use of the correlation between variables: if $V_i$ denotes the set of variables which have been observed for the $i$th object, one might expect the values of $\{x_{ik} \ (k \in V_i)\}$ to provide information relevant for estimating the missing value $x_{im}$. For quantitative variables, one can estimate $x_{im}$ by (Buck, 1960)

$$\hat{x}_{im} \equiv \bar{x}_{.m} + \Sigma_{k \in V_i} b_{mk}(x_{ik} - \bar{x}_{.k}),$$

where

$$\bar{x}_{.k} \equiv \text{mean}\{x_{jk} \mid j \in C_i\},$$

and $\{b_{mk} \ (k \in V_i)\}$ are obtained from the minimization of

$$\Sigma_{j \in C_i}\{x_{jm} - \bar{x}_{.m} - \Sigma_{k \in V_i} b_{mk}(x_{jk} - \bar{x}_{.k})\}^2.$$

Little and Rubin (1987, Chapter 12) discuss other models relevant for the estimation of missing values for various types of variable. In practice, several different imputed values for $x_{im}$ would be obtained, by varying the method of imputation and the specification of the set $C_i$ of similar objects; if similar imputed values were obtained, one could have more confidence in the results. An example is provided in Gordon (1981, Section 7.4).

### 2.3.3 Conditionally-present variables

It is sometimes the case that some of the variables are not relevant for all of the objects under investigation. For example, properties of the petals of plants and the wings of insects are defined conditional on the binary variables 'presence of petals' and 'presence of wings' belonging to the state 'present'. Variables like 'colour of petals' or 'petal size', which might not be recordable for some of the objects under study, are examples of conditionally-present or serially dependent (Williams, 1976, Chapter 5) variables.

It is important to ensure that the contributions of such variables to an overall measure of pairwise (dis)similarity are suitably weighted (Kendrick, 1965). Consider three species of plant, $A$, $B$ and $C$, of which only $A$ and $B$ possess petals. The fact that $A$ and $B$ differ from one another in the properties of their petals should not of itself make either of them more similar to $C$ than they are to each other.

This requirement is met by the general (dis)similarity coefficient M10. Assume that the contribution provided by any variable to the overall measure of dissimilarity is bounded above by 1. If variables $k_1, k_2, ..., k_r$ can be recorded only if the $k$th variable takes the state 'present', define

$$d_{ijk} \equiv \Sigma_{t=1}^{r} w_{ijk_t} d_{ijk_t} / \Sigma_{t=1}^{r} w_{ijk_t}$$

for objects $i$ and $j$ for which information is available on the conditionally-present variables $k_1, k_2, ..., k_r$. If the $k$th variable is 'present' for only one of objects $i$ and $j$, $d_{ijk}$ is defined to be 1; if neither of the objects has the $k$th variable in the state 'present', $d_{ijk}$ is defined to be 0. This ensures that $d_{ijk}$ is bounded by 0 and 1. When constructing the overall measure of dissimilarity M10 from $\{d_{ijk}$ $(k = 1, ..., p)\}$, conditionally-present variables are not included in the summation.

## 2.4 Construction of relevant measures

Section 2.2 described some of the ways in which the information contained in a pattern matrix can be transformed into a matrix of similarities $(s_{ij})$ or dissimilarities $(d_{ij})$ between each pair of objects. Many other definitions are possible. On occasion, investigators may be able to specify measures that are particularly relevant for their data sets, but the selection of appropriate measures of (dis)similarity is often far from straightforward. This section presents some comments, mostly formulated as questions for consideration, which are intended to help investigators select measures that are relevant for their data sets.

1. The measures described in Section 2.2 are categorized by the types of variable used in the description of the objects but has the appropriate type of variable been used in the pattern matrix? For example, the fact that a variable can be measured on a continuous scale does not necessarily mean that it should be retained as a quantitative variable: external information about the objects may indicate that it is more relevant to record it as a binary variable with states defined by whether or not the value lies above a pre-specified threshold value.

2. For binary variables, how should 'co-absence' be regarded, i.e. if a variable belongs to the state '−' or 'absent' for each of two objects, should this increase the similarity between the objects? The answer depends on the status of the two states of the variable. In Example 2, the absence of a plant species from an ecological quadrat is qualitatively different from its presence, and one would usually want to ignore co-absences in constructing measures of (dis)similarity and use measures such as M2. However, if the binary variable recorded the gender of an individual, the states 'male' and 'female' are genuine alternatives and one would want to use a measure such as M1 that takes account of both types of matching.

3. For quantitative variables, is one interested in differences between the actual values taken by the variables or in the relative magnitude of the set of variables describing the two objects? In the latter case, interest would appear to be concentrated on the 'shape' of objects, rather than their 'size', and measures such as M8 and M9 may be more appropriate. However, before using the correlation coefficient M9, the relevance of the concept of the mean value $\bar{x}_{i\cdot}$ of an 'object' should be assessed for the data under in-

vestigation. If differences between the values of variables are of interest, measures M4 – M7 are indicated as appropriate.

4. Consider two objects whose values differ by the same amount on two commensurate quantitative random variables, i.e.

$$\mid x_{ik} - x_{jk} \mid \; = \; \mid x_{il} - x_{jl} \mid .$$

Should the contribution of these two differences to a measure of the dissimilarity of the $i$th and $j$th objects be ($i$) the same, or ($ii$) dependent on the relative magnitudes of $(x_{ik}, x_{jk})$ and $(x_{il}, x_{jl})$, or ($iii$) dependent on the relative magnitudes of the amount of variability displayed by the $k$th and $l$th variables? In case ($i$), an unweighted Minkowski metric (M4 – M6 with $w_k = 1 \; (k = 1, ..., p)$) is indicated as appropriate. In case ($ii$), the Canberra metric M7 would scale down the contribution from larger values, as would square root transformations of the $x_{ik}$'s; transformations to scale up the effect of larger values would seem to have limited applicability. In case ($iii$), use of a Minkowski metric with suitably chosen weights $\{w_k \; (k = 1, ..., p)\}$ may be appropriate, e.g. $w_k^{-1}$ could be chosen to be the $k$th variable's standard deviation or range of values in the set of objects.

5. For quantitative variables in cases ($i$) and ($iii$) above, how strongly should differences between variables contribute to the measure of dissimilarity? The Minkowski parameter $\lambda$ in M4 allows various powers of the differences to be used, although the desire for simplicity and the difficulty of justifying more general values may prevent investigators from making choices other than $\lambda = 1$ or $\lambda = 2$. The possibility that some power transformation of the variables may be more appropriate should be considered.

Investigators may find it helpful to contemplate how they would quantify the dissimilarities between selected (possibly hypothetical) objects. A more formal procedure for modelling dissimilarities and estimating appropriate weightings of variables may be feasible in some instances (Sokal and Rohlf, 1980; Gordon, 1990). In this approach, a measure $d_{ijk}$ is defined of the partial contribution of the $k$th variable to the dissimilarity between the $i$th and $j$th objects, e.g. $d_{ijk} \; = \; \mid x_{ik} - x_{jk} \mid^{\lambda_k}$ for a quantitative variable. The overall dissimilarity between the $i$th and $j$th objects is defined to be a weighted average of $(d_{ijk} \; (k = 1, ..., p))$:

$$d_{ij}(\mathbf{w}) \equiv \Sigma_{k=1}^{p} w_k d_{ijk},$$

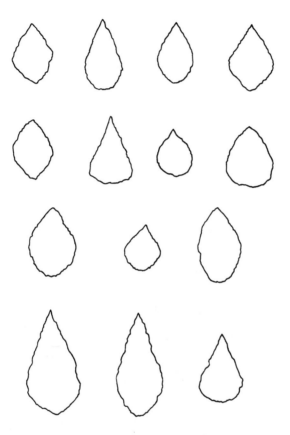

Figure 2.2 *Outlines of 14 flint arrowheads from north-east Scotland. Reproduced from Gordon (1990).*

where
$$w_k \geq 0 \ (k = 1, ..., p)$$
and
$$\Sigma_{k=1}^{p} w_k = 1.$$

For a small number of object pairs $(i, j)$, a measure of the perceived dissimilarity $\delta_{ij}$ is elicited directly from an assessor and values of the weights $\mathbf{w} \equiv \{w_k \ (k = 1, ..., p)\}$ are sought that ensure that $(d_{ij}(\mathbf{w}))$ provides a 'good' approximation to $(\delta_{ij})$. Two

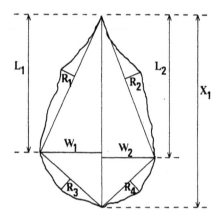

Figure 2.3 *Measurements made on flint arrowheads; see also Table 2.1. Reproduced from Gordon (1990).*

Table 2.1 *Definition of variables characterizing the flint arrowheads; see also Fig. 2.3*

---

$X_1 \equiv$ maximum length of arrowhead
$X_2 \equiv (W_1 + W_2)/X_1$, ratio of maximum width to maximum length
$X_3 \equiv (L_1 + L_2)/X_1$, location of maximum width
$X_4 \equiv (R_1 + R_2)/X_1$, convexity of upper part of arrowhead
$X_5 \equiv (R_3 + R_4)/X_1$, convexity of lower part of arrowhead

---

models relating $(\delta_{ij})$ and $(d_{ij}(\mathbf{w}))$ that have been investigated involve linking them using a linear or a monotone function. In the linear model,

$$\delta_{ij} \approx c d_{ij}(\mathbf{w}) \quad (c \geq 0),$$

where '$\approx$' denotes 'is approximately equal to'. This model is fitted using least squares multiple linear regression with no intercept term. In the monotone model,

$$\delta_{ij} \approx M(d_{ij}(\mathbf{w})),$$

where $M$ denotes a monotone function. This model is fitted using Kruskal's (1964a, b) method of least squares monotone regression with primary treatment of ties, a description of which is presented in Section 6.3.

Table 2.2 *Optimal weights of the five variables characterizing the flint arrowheads*

| Model | Estimates of | | | | |
|-------|------|------|------|------|------|
|       | $w_1$ | $w_2$ | $w_3$ | $w_4$ | $w_5$ |
| Linear | 0.15 | 0.27 | 0.32 | 0.20 | 0.06 |
| Monotone | 0.17 | 0.35 | 0.28 | 0.17 | 0.03 |

This methodology is illustrated by application to the flint arrowheads data. Fig. 2.2 depicts the outlines of the 14 flint arrowheads, which are a subset selected from a larger collection of arrowheads. In order to classify this larger set of arrowheads, it would be helpful to define a measure of their pairwise dissimilarity. Table 2.1 and Fig. 2.3 define five variables which were believed to be relevant for distinguishing between the arrowheads, the first of which provides a measure of the size of an arrowhead and the other four of which provide measures of its shape. Each of these variables was standardized to have unit variance; although not strictly necessary, this transformation has the effect of making the weights $\{w_k$ $(k = 1, ..., 5)\}$ more nearly equal. The partial contribution $d_{ijk}$ of the $k$th variable towards the dissimilarity between the $i$th and $j$th arrowheads was defined by

$$d_{ijk} = (x_{ik} - x_{jk})^2 \ (1 \leq j < i \leq 14; \ k = 1, ..., 5).$$

An assessor provided his opinion of the overall dissimilarities $(\delta_{ij})$ between each of the 91 pairs of arrowheads, and the weights $\{w_k$ $(k = 1, ..., 5)\}$ which ensured that $(d_{ij}(\mathbf{w}))$ provided an optimal fit to $(\delta_{ij})$ under the linear and monotone models are given in Table 2.2. Very similar results were obtained under the two models: in each case, $w_5$ is close to zero and the other four variables are weighted approximately in the ratios 1:2:2:1. After suitable standardization of variables, these weights could be used to construct a dissimilarity matrix for the larger collection of arrowheads. However, it should be noted that several decisions were taken during the construction of the model, e.g. the definition of $(d_{ijk})$ and the specification of the set of variables to be included; it is possible that other decisions would have allowed a more accurate modelling of the assessor's judgments. Further discussion of such problems is given by Gordon (1990).

The construction of relevant measures of pairwise (dis)similarity can be quite challenging, but the quality of a classification can be enhanced if investigators are prepared to devote a reasonable amount of time to the matter. If there is uncertainty about some of the decisions that are taken, a small number of different measures can be evaluated and analysed in order to establish whether or not they would lead to markedly different conclusions about the data.

CHAPTER 3

# Partitions

## 3.1 Partitioning criteria

This chapter addresses the problem of partitioning a set of $n$ objects into $c$ disjoint classes. It is assumed in the first four sections that the value of $c$ is known; comments on ways of determining appropriate values of $c$ are presented in Section 3.5.

The aim is to find a partition in which objects are similar to the other objects belonging to their class and dissimilar to objects that belong to different classes. A general approach involves defining a measure of the adequacy of a partition and seeking a partition of the objects which optimizes that measure. Several possible measures of adequacy are defined in this section. It is convenient to start by defining, for a class of objects, measures of

- its heterogeneity, or lack of cohesion

- its isolation or separation from the rest of the data.

Some of these measures require the objects to be described by a set of quantitative variables: thus, $x_{ik}$ denotes the value of the $k$th variable describing the $i$th object $(i = 1, ..., n; k = 1, ..., p)$. Other measures are based solely on a matrix of pairwise dissimilarities $(d_{ij})$, where $d_{ij}$ denotes the dissimilarity between the $i$th and $j$th objects $(i, j = 1, ..., n)$. The measures are illustrated by application to a set of seven objects described by two variables, as recorded in Table 3.1; a plot of the objects on these variables is shown in Fig. 3.1. Table 3.2 presents the *squared* Euclidean distances between the objects, these squared distances being regarded as a measure of pairwise dissimilarity in the illustration of several of the measures. The values of the measures for these data are given in Table 3.3.

Five measures of the heterogeneity $H(C_r)$ of a class $C_r$ are defined below.

Table 3.1 *Description of seven objects in terms of two quantitative variables.*

| Object, $i$ | 1 | 2 | 3 | 4 | 5 | 6 | 7 |
|---|---|---|---|---|---|---|---|
| $x_{i1}$ | 9 | 18 | 24 | 25 | 32 | 34 | 40 |
| $x_{i2}$ | 33 | 7 | 23 | 40 | 47 | 30 | 16 |

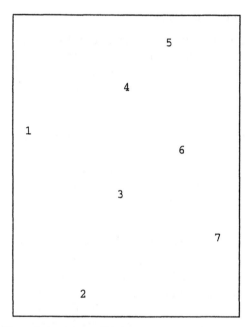

Figure 3.1 *Plot of the seven objects whose coordinate values are given in Table 3.1.*

*H1: Sum of squares*

$$H(C_r) \equiv \Sigma_{i \in C_r} \Sigma_{k=1}^{p} (x_{ik} - \bar{x}_{.k}^{(r)})^2,$$

where

$$\bar{x}_{.k}^{(r)} \equiv \text{mean}\{x_{ik} \mid i \in C_r\} \ (k = 1, ..., p).$$

An algebraically-equivalent expression is

$$H(C_r) \equiv \Sigma_{i,j \in C_r} d_{ij}/(2n_r),$$

Table 3.2 *Squared Euclidean distances between the objects described in Table 3.1.*

| Object | | | | | | |
|--------|------|------|-----|-----|------|-----|
| 2 | 757 | | | | | |
| 3 | 325 | 292 | | | | |
| 4 | 305 | 1138 | 290 | | | |
| 5 | 725 | 1796 | 640 | 98 | | |
| 6 | 634 | 785 | 149 | 181 | 293 | |
| 7 | 1250 | 565 | 305 | 801 | 1025 | 232 |
| Object | 1 | 2 | 3 | 4 | 5 | 6 |

where $d_{ij}$ denotes the *squared* Euclidean distance between the $i$th and $j$th objects, and $n_r$ denotes the number of objects belonging to class $C_r$ (Edwards and Cavalli-Sforza, 1965).

The sum of squares of the cluster comprising the seven objects described in Tables 3.1 and 3.2 is 1798, as may be verified either by evaluating the sum of the squared distances of the objects from the class centroid (26, 28), or by dividing by 7 the sum of the elements of the lower triangular matrix given in Table 3.2.

*H2: $L_1$ measure*

$$H(C_r) \equiv \Sigma_{i \in C_r} \Sigma_{k=1}^p \mid x_{ik} - m_{.k}^{(r)} \mid,$$

where

$$m_{.k}^{(r)} \equiv \text{median}\{x_{ik} \mid i \in C_r\} \ (k = 1, ..., p).$$

The 'median' of the seven objects is (25, 30), the sum of the $L_1$ distances about this point being 129.

*H3: Diameter*

$$H(C_r) \equiv \max_{\{i,j \in C_r\}} d_{ij},$$

the dissimilarity between the most dissimilar pair of objects, in this case the second and fifth objects.

*H4: Star*

$$H(C_r) \equiv \min_{\{i \in C_r\}} \Sigma_{j \in C_r} d_{ij},$$

the name arising from the graph formed by linking together all pairs of objects whose pairwise dissimilarities are included in the sum. The smallest sum occurs when the third object is located at the centre of the star.

*H5: Sum of distances*

$$H(C_r) \equiv \Sigma_{\{i,j \in C_r | j < i\}} d_{ij},$$

the sum of all the elements of the lower triangular matrix.

Three of these measures of class heterogeneity include the definition of a 'centre' of the class: the object at the centre of the star in *H4* and (if the objects are described by quantitative variables) the mean in *H1* and the 'median' in *H2*. These centres provide simple summaries of each class; more elaborate descriptions are presented in Section 7.3.

Two measures of the isolation $I(C_r)$ of class $C_r$ are defined below. For illustrative purposes, the measures are evaluated for class $\{3, 6\}$.

*I1: Split*

$$I(C_r) \equiv \min_{\{i \in C_r, j \notin C_r\}} d_{ij},$$

the smallest dissimilarity between an object in the class and an object outside the class. For class $\{3, 6\}$, this is the dissimilarity between objects 4 and 6.

*I2: Cut*

$$I(C_r) \equiv \Sigma_{i \in C_r} \Sigma_{j \notin C_r} d_{ij},$$

the sum of the dissimilarities between objects in the class and objects outside the class.

Other measures of the heterogeneity and isolation of a class are described by Hansen and Jaumard (1997).

Given measures of the heterogeneity or isolation of each class in a partition, these can be combined to provide measures of the

Table 3.3 *Values taken by measures of heterogeneity and isolation for the data described in Tables 3.1 and 3.2.*

| H1 | 1798 | H3 | 1796 | I1 | 181 |
|----|------|----|------|----|------|
| H2 | 129 | H4 | 2001 | I2 | 3977 |
| | | H5 | 12586 | | |

adequacy of the partition. Such measures can be based on the value of the 'worst' class or on the average (or sum) of the values over all classes. Optimal partitions are those which minimize measures of heterogeneity, maximize measures of isolation, or are based on a combination of these desiderata; of course, there can be more than one partition which takes the optimal value of a given criterion.

Two general families of heterogeneity criteria for the adequacy of a partition into the c classes, $\{C_1, C_2, ..., C_c\}$, are

$$P(H, \Sigma) \equiv \Sigma_{r=1}^{c} H(C_r)$$

and

$$P(H, Max) \equiv \max_{\{r=1,...,c\}} H(C_r).$$

For example, use of criterion $P(H1, \Sigma)$ requires seeking a partition with minimum total sum of squares, whereas criterion $P(H3, Max)$ defines an optimal partition to be one for which the largest diameter of one of its constituent classes is as small as possible.

Two general families of isolation criteria for the adequacy of a partition into the c classes, $\{C_1, C_2, ..., C_c\}$, are

$$P(I, \Sigma) \equiv \Sigma_{r=1}^{c} I(C_r)$$

and

$$P(I, min) \equiv \min_{\{r=1,...,c\}} I(C_r).$$

For example, criterion $P(I1, min)$ defines an optimal partition to be one for which the smallest split of its constituent classes is as large as possible.

There are links between some criteria: thus, minimizing $P(H5, \Sigma)$ is equivalent to maximizing $P(I2, \Sigma)$.

Partition heterogeneity criteria and partition isolation criteria can be perceived as concentrating on maximizing, respectively, the cohesion and the isolation of the clusters in the partition. However, this statement requires qualification: in addition to the link between $P(H5, \Sigma)$ and $P(I2, \Sigma)$ noted above, the total sum of squares of a set of points about their centroid can be partitioned into within-class and between-class components,

$$
\begin{aligned}
\Sigma_{r=1}^c \Sigma_{i \in C_r} \Sigma_{k=1}^p (x_{ik} - \bar{x}_{.k})^2 \;=\; & \Sigma_{r=1}^c \Sigma_{i \in C_r} \Sigma_{k=1}^p (x_{ik} - \bar{x}_{.k}^{(r)})^2 \\
& + \; \Sigma_{r=1}^c \Sigma_{k=1}^p n_r (\bar{x}_{.k}^{(r)} - \bar{x}_{.k})^2,
\end{aligned}
$$

where

$$
\bar{x}_{.k} \equiv \Sigma_{r=1}^c \Sigma_{i \in C_r} x_{ik} / \Sigma_{r=1}^c n_r \quad (k = 1, ..., p).
$$

The first term on the right-hand side of the equation is $P(H1, \Sigma)$, a measure of the heterogeneity of the classes in the partition, and the second term may be regarded as a measure of the isolation of the classes in the partition; hence, in this case, minimizing heterogeneity is equivalent to maximizing isolation.

The problem of finding a partition of a set of objects into $c$ classes which optimizes a stated criterion of partition adequacy is not, in general, straightforward. The number of different ways, $N(n, c)$, in which $n$ objects can be partitioned into $c$ non-empty classes is

$$
N(n, c) \;=\; \frac{1}{c!} \sum_{k=1}^c (-1)^{c-k} \binom{c}{k} k^n
$$

(Jensen, 1969). As $n$ and $c$ increase, it soon becomes computationally infeasible to examine all possible partitions in order to identify one with optimal value of a stated criterion; for example, $N(19, 8) > 1.7 \times 10^{12}$. For some criteria (e.g. $P(I1, min)$, see Section 3.4.1), an optimal partition can be identified without searching over the set of partitions. For other criteria, efficient accounting schemes based on dynamic programming or branch-and-bound algorithms can reduce the number of partitions that require to be examined (Jensen, 1969; Koontz, Narendra and Fukunada, 1975; Diehr, 1985), but even with such savings a global search makes too heavy a demand on computing facilities for even moderate-sized values of $n$.

In fact, many partitioning problems are NP-hard: crudely, there are no known algorithms for solving them which require a number of operations that is polynomial in $n$, and expert opinion considers it unlikely that any polynomial-time algorithms exist (Garey and

Johnson, 1979; Day, 1996). There is thus a need for efficient algorithms for obtaining partitions, where the definition of the word 'efficient' is taken to mean that the algorithm not only provides a partition whose criterion value is close to optimal, but also makes minimal demands on computing resources. The next three sections describe algorithms that have been used to find partitions.

## 3.2 Iterative relocation algorithms

Given a criterion measuring the adequacy of a partition of a set of objects into $c$ disjoint classes, there are two main stages in iterative relocation algorithms that search for 'optimal' partitions of the objects. These involve the specification of:

- an initial partition of the objects into $c$ classes

- a set of transformations for changing a partition into another partition.

The partition is then modified until none of the allowable transformations would improve the given criterion of partition adequacy.

### 3.2.1 Initial partition

This can be obtained by assigning objects to classes randomly (MacQueen, 1967) or using the results of a previous investigation, for example by sectioning a hierarchical classification to provide a partition of the objects into $c$ classes or by amalgamating a 'closest' pair of classes in a partition of the objects into $(c + 1)$ classes (Beale, 1969; Wishart, 1987).

Alternatively, if the set of objects can be represented as a set of points in some space, $c$ seed points can be specified and each object assigned to a seed point that is closest to it. These seed points can be chosen evenly over the region of space in which the set of objects is located, or can comprise a subset of the objects. In this latter case, the objects can be chosen sequentially such that they are:

- greater than a specified distance from any object that has already been specified as a seed point

- farthest from any of the existing seed points

- selected from high-density regions of space

The first strategy has also been used on its own, without any relocation stage, as a 'single pass' algorithm for the analysis of large data sets. By contrast, the second and third strategies require consideration of the entire data set before the seeds are selected.

### 3.2.2 Transformations

One of the earliest iterative relocation algorithms involved itera- tively modifying a partition by simultaneously assigning each ob- ject to a class to whose centroid it was closest (tied distances being broken arbitrarily) and then recalculating class centroids (e.g. Ball and Hall, 1967). Research based on this algorithm did not always explicitly state a criterion of partition adequacy to be optimized, but it can be shown that (in the absence of tied centroid-to-object distances) the modifications made during an iteration of the al- gorithm reduce the sum of squares criterion $P(H1, \Sigma)$ defined in the previous section. However, this criterion would also be reduced if the $i$th point were moved from the $r$th class to the $s$th class whenever

$$D_{is}^2 < [(n_s + 1)n_r/(n_s(n_r - 1))]D_{ir}^2,$$

where $D_{ir}$ (resp., $D_{is}$) denotes the distance of the $i$th object from the centroid of the $r$th (resp., $s$th) class, which contains $n_r$ (resp., $n_s$) objects (Beale, 1969).

Instead of assigning all objects simultaneously to a nearest class centroid, partitions can alternatively be modified by considering relocating a single object at a time, this approach also being appli- cable for other definitions of the heterogeneity of a partition. Many different ways of relocating a single object have been investigated (e.g. Friedman and Rubin, 1967; MacQueen, 1967; Hartigan and Wong, 1979; Ismail and Kamel, 1989): the objects can be investi- gated for possible relocation in a random or systematic order; the object chosen for relocation can be the first one encountered whose movement would reduce the given criterion of partition hetero- geneity or the one which would most improve the criterion value; the object can be moved to the first different class encountered which would improve the quality of the partition or to the class which would yield maximum improvement, or a hybrid of these two strategies can be adopted; class centroids can be updated af- ter each relocation or only after all $n$ objects have been considered for relocation. A more elaborate transformation involves the pairwise interchange of the class memberships of two objects (Banfield and Bassill, 1977).

Such algorithms, involving the identification of a centroid for each class, are often referred to as $k$-means or $c$-means algorithms. When used to address the problem of the efficient coding of signals, the methodology has also been called 'vector quantization' (Linde, Buzo and Gray, 1980; Lloyd, 1982).

The algorithm runs until the quality of the partition cannot be further improved using the available set of transformations; this provides a local optimum (which need not be a global optimum) of the criterion of partition adequacy. Clearly, a larger set of allowable transformations would make it more likely that the final partition provided a global optimum of the criterion of partition adequacy, but the expected improvement is paid for by the increased cost of computation.

Simulated annealing (Kirkpatrick, Gelatt and Vecchi, 1983) algorithms have also been investigated: in these algorithms, a transformation which would degrade the quality of a partition is not ruled out but has a small probability of being implemented; the aim is to prevent the algorithm becoming trapped in an inferior local optimum solution (Klein and Dubes, 1989; Selim and Asultan, 1991; Sun et al., 1994). The success of such algorithms depends heavily on the parameters of the cooling schedule and they have not yet been widely used to address classification problems.

Parallel computer hardware (Hwang and Briggs, 1984) is also potentially valuable for obtaining partitions, particularly when an algorithm involves repeated application of an operation, such as evaluating distances between objects and current class centroids. Typically, a unique processing element is assigned to each element of the pattern matrix. Applications include iterative relocation algorithms seeking a partition which minimizes $P(H1, \Sigma)$ (Ni and Jain, 1985; Li and Fang, 1989) and a 'single pass' algorithm (Pogue, Rasmussen and Willett, 1988; Whaley and Hodes, 1991).

There are several extensions of the basic iterative relocation algorithm. It is possible to allow a limited number of 'atypical' objects to be placed in a 'residue' class which is disregarded when the criterion of partition adequacy is evaluated (Wishart, 1987) or for the number of classes, $c$, to be changed during the course of the algorithm (Ball and Hall, 1967; MacQueen, 1967): thus, classes which are regarded as too heterogeneous can be subdivided and those

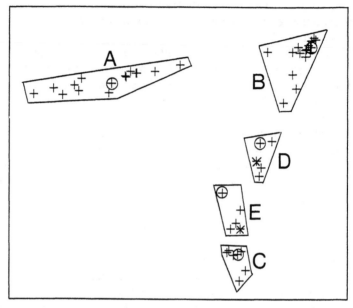

Figure 3.2 *A plot of the Abernethy Forest 1974 data on their first two principal components, showing a partition of the data into five classes.*

whose centres are perceived as too close together can be amalgamated; however, this requires various threshold parameters to be specified by the investigator. Additionally, once an (approximation to an) optimal partition into $c$ classes has been found, the 'most similar' pair of classes can be amalgamated and the procedure repeated for partitions into successively smaller numbers of classes (Beale, 1969; Wishart, 1987), to yield a hybrid between an iterative relocation algorithm and an agglomerative algorithm.

Iterative relocation algorithms should be run several times, with different initial partitions in each run. If the best partition obtained is found a high proportion of the time, one can have more confidence that this is the globally optimal partition with respect to the stated criterion of partition adequacy.

### 3.2.3 Examples

Fig. 3.2 shows a graphical representation of the Abernethy Forest 1974 data. The Figure was obtained using principal components

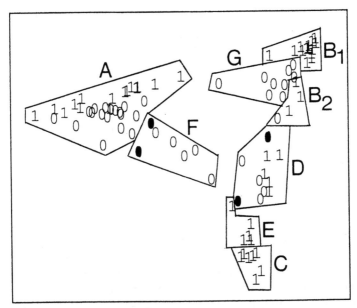

Figure 3.3 *A plot of the combined Abernethy Forest data sets on their first two principal components, showing a partition of the data into eight classes.*

analysis (of the covariance matrix: see Section 6.2) to reduce the nine-dimensional data to a two-dimensional approximation, but it should be noted that the analyses which follow are based on the original nine variables. Methodology presented in Section 3.5 suggests that there are five classes in this data set, and Fig. 3.2 also shows the partition of the data into five classes obtained by minimizing $P(H1, \Sigma)$ using an iterative relocation algorithm; for these data, several different initial partitions all provided the same final partition shown in Fig. 3.2. The class structure appears quite clear for these data, with the classes A – E shown in Fig. 3.2 comprising the following samples:

$$A: \{1\text{--}15\}; \ B: \{16\text{--}32\}; \ C: \{35\text{--}41\};$$
$$D: \{33, 42\text{--}45\}; \ E: \{34, 46\text{--}49\}.$$

Apart from samples 33 and 34, which are shown starred, the partition respects the stratigraphical ordering of the samples. The explanation of the circled points is postponed to Section 3.3.

Table 3.4 *Membership of the eight classes into which the combined Abernethy Forest data set is partitioned in Fig. 3.3.*

| Class | Abernethy Forest 1974 data | Abernethy Forest 1970 data |
|-------|----------------------------|----------------------------|
| $A$   | 1 – 15                     | 1, 3 – 17, 19, 23          |
| $B_1$ | 16 – 29                    | –                          |
| $B_2$ | 30 – 32                    | 32                         |
| $C$   | 35 – 41                    | –                          |
| $D$   | 33, 42 – 45                | 33, 35, 37 – 40            |
| $E$   | 34, 46 – 49                | –                          |
| $F$   | –                          | 2, 18, 20 – 22, 36, 41     |
| $G$   | –                          | 24 – 31, 34                |

Fig. 3.3 shows a plot on the first two principal components of the combined Abernethy Forest data sets, with samples from the 1970 data set being labelled '0' and samples from the 1974 data set being labelled '1'. Also shown is the partition obtained when an iterative relocation algorithm was used to seek a partition into eight classes which minimizes $P(H1, \Sigma)$. The displayed partition, which might not have overall smallest value of $P(H1, \Sigma)$, differs from the partition obtained by sectioning the corresponding hierarchical classification in the placement of four samples from the 1970 data set (shown by filled zeros).

The 1974 samples are located in similar positions in Figs. 3.2 and 3.3, and their classification in the combined analysis corresponds to their earlier classification with the exception that class $B$ is subdivided into classes $B_1$ and $B_2$. However, the classification of the combined data set is less clear-cut than that of the 1974 data set on its own, and not all of the classes contain samples from both data sets (see Table 3.4), suggesting differences over time in the composition of the vegetation surrounding the two sites.

Diday and Govaert's data, plotted in Fig. 3.4, highlight a drawback of the sum of squares criterion $P(H1, \Sigma)$. A partition of these data into three classes which minimized $P(H1, \Sigma)$ was sought, using the iterative relocation algorithm. The optimal partition that was found, in which 23 of the 150 objects were assigned to a class other than the one specified by their parent distribution, is shown in Fig. 3.5. This Figure illustrates the tendency of the sum of squares criterion to produce classes that occupy hyperspherical re-

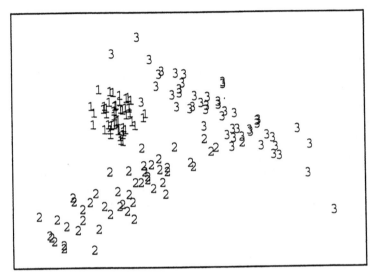

Figure 3.4 *A plot of Diday and Govaert's data, comprising samples of size 50 from each of three bivariate normal distributions; the labels indicate the provenance of each datum.*

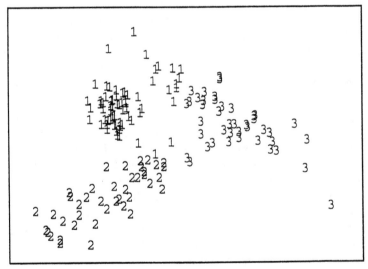

Figure 3.5 *A sum of squares partition into three classes of Diday and Govaert's data.*

gions of space. Other classification criteria also involve such implicit models for the type of class structure that is assumed to be present in a data set; a further discussion of this topic is presented in Section 3.6.

Given the wide range of different types of structure that could be present in a data set, one would ideally prefer to use classification procedures which 'adapted' themselves to the type of data which they were investigating. A basic idea common to such adaptive procedures is that the measure of distance from, or variability within, a class of objects depends on the configuration of the objects within that class. For example, if $S_r$ denotes the sample covariance matrix for the $r$th class, the squared distance from a point $y_r$ in that class to some other point $x$ can be defined to be the Mahalanobis distance,

$$\Delta(x, y_r) \equiv (y_r - x)' S_r^{-1} (y_r - x),$$

or to be a standardized version of the Mahalanobis distance which inhibits the growth of classes occupying large regions of space, e.g.

$$\Delta_1(x, y_r) \equiv \{\det(S_r)\}^{1/p} \Delta(x, y_r),$$

where $p$ denotes the number of variables describing the objects.

These measures of distance are not symmetric, but depend on the class with respect to which they are being evaluated. Diday and Govaert (1977) describe an adaptive generalization of the $c$-means algorithm. In each iteration of this algorithm, each object is assigned as before to a class to which it is closest, but in this case the distance of the $i$th object $x_i$ from the centroid $y_r$ of the $r$th class is measured by $\Delta_1(x_i, y_r)$. The optimal partition found when this algorithm was applied to Diday and Govaert's data is shown in Fig. 3.6; this partition was obtained from several different random initial partitions and also when the initial partition was specified by the 'correct' partition shown in Fig. 3.4. The partition shown in Fig. 3.6 assigns five points to a class other than the one specified by their parent distribution. The 'misclassified' points are circled in Fig. 3.6, from which it is clear that at least two of them are assigned to their natural classes.

Other adaptive classification procedures have been proposed by Lefkovitch (1978, 1980) and Art, Gnanadesikan and Kettenring (1982).

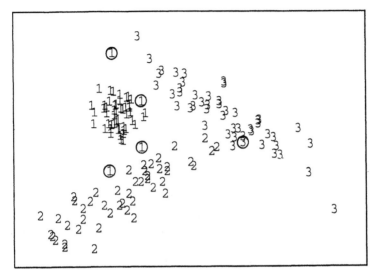

Figure 3.6 *An adaptive classification of Diday and Govaert's data, with 'misclassified' points shown circled.*

## 3.3 Mathematical programming

A partition of $n$ objects can be defined in several distinct ways by specifying a set of 0/1 variables. Three different sets of such variables are defined below; in each case, some side conditions are required to ensure that they define a partition into the requisite number of classes.

(i) $y_{ir} = 1$ (resp., 0) if the $i$th object belongs (resp., does not belong) to the $r$th class ($i = 1, ..., n; r = 1, ..., c$)

(ii) $z_{ij} = 1$ (resp., 0) if the $i$th and $j$th objects belong (resp., do not belong) to the same class ($i, j = 1, ..., n$)

(iii) $v_r = 1$ (resp., 0) if the $r$th class belongs (resp., does not belong) to the partition ($r = 1, ..., 2^n - 1$)

The problems of finding such sets of 0/1 variables which specify optimal partitions can be formulated as mathematical programming problems. Three examples are given below.

First, the problem of minimizing the sum of within-class distances, $P(H5, \Sigma)$, can be expressed as (Klein and Aronson, 1991):

MP1: minimize

$$\Sigma_{i=2}^{n}\Sigma_{j=1}^{i-1}d_{ij}z_{ij}$$

subject to

$$z_{ij} \geq y_{ir} + y_{jr} - 1 \ (1 \leq j < i \leq n; r = 1,...,c) \qquad (3.1)$$

$$\Sigma_{r=1}^{c}y_{ir} = 1 \ (i = 1,...,n) \qquad (3.2)$$

$$y_{ir} \in \{0,1\} \ (i = 1,...,n; r = 1,...,c) \qquad (3.3)$$

$$z_{ij} \geq 0 \ (1 \leq j < i \leq n). \qquad (3.4)$$

The side conditions (3.1) ensure that the values of $(y_{ir})$ and $(z_{ij})$ do not provide contradictory information. Conditions (3.2) require each object to belong to precisely one of the $c$ classes. It is not necessary to specify that each $z_{ij}$ be either 0 or 1 as this is enforced by the objective function and other conditions.

Secondly, the problem of minimizing the sum of the star distances, $P(H_4, \Sigma)$, can be expressed as (Vinod, 1969; Mulvey and Crowder, 1979; Massart, Plastria and Kaufman, 1983):

MP2: minimize

$$\Sigma_{i=1}^{n}\Sigma_{j=1}^{n}d_{ij}y_{ij}$$

subject to

$$y_{ij} \leq y_{jj} \ (i, j = 1,...,n) \qquad (3.5)$$

$$\Sigma_{j=1}^{n}y_{ij} = 1 \ (i = 1,...,n) \qquad (3.6)$$

$$\Sigma_{j=1}^{n}y_{jj} = c \qquad (3.7)$$

$$y_{ij} \in \{0,1\} \ (i, j = 1,...,n). \qquad (3.8)$$

In this formulation, the definition of the set $(y_{ij})$ is somewhat different to that given previously: $y_{ij} = 1$ (resp., 0) if the $i$th object belongs (resp., does not belong) to the class whose star centre (or 'medoid': Kaufman and Rousseeuw (1990)) is the $j$th object $(i, j = 1,...,n)$; clearly, $y_{jj} = 1$ (resp., 0) if the $j$th object is selected (resp., not selected) as one of the medoids $(j = 1,...,n)$.

Conditions (3.5) require $y_{jj}$ to take the value 1 if the $j$th object is the medoid of a class, conditions (3.6) ensure that each object belongs to precisely one class, and condition (3.7) specifies the number of classes to be $c$.

This problem is the 'optimal facility location' (also called the 'c-median') problem, in which $c$ facilities have to be located at some of $n$ specified sites so as to minimize the sum of the distances from each site to its nearest facility.

Thirdly, one can implicitly consider including in the partition any of the $2^n - 1$ possible non-empty classes (Hansen and Jaumard, 1997):

MP3: minimize

$$\Sigma_{r=1}^{2^n-1} H(C_r) v_r$$

subject to

$$\Sigma_{r=1}^{2^n-1} a_{ir} v_r = 1 \ (i = 1, ..., n) \tag{3.9}$$

$$\Sigma_{r=1}^{2^n-1} v_r = c \tag{3.10}$$

$$v_r \in \{0, 1\} \ (r = 1, ..., 2^n - 1), \tag{3.11}$$

where $H(C_r)$ denotes a measure of the heterogeneity of the $r$th class $(r = 1, ..., 2^n - 1)$, and $a_{ir} = 1$ (resp., 0) if the $i$th object belongs (resp., does not belong) to the $r$th class $(i = 1, ..., n; r = 1, ..., 2^n - 1)$; this latter notation is used to stress the fact that obtaining the values of $(a_{ir})$ is not part of the minimization problem. Conditions (3.9) ensure that each object belongs to precisely one of the classes included in the partition, and condition (3.10) specifies the number of classes in the partition to be $c$.

A wide range of approaches has been used to provide solutions to these mathematical programming problems. MP1 and MP2 are (mixed) integer linear programming problems. The continuous relaxation of these problems involves replacing integer conditions by inequality conditions, e.g. conditions (3.3) would be replaced by:

$$0 \leq y_{ir} \leq 1 \ (i = 1, ..., n; r = 1, ..., c), \tag{3.12}$$

to yield a linear programming problem. If the solution to the linear program has all variables taking values 0 or 1, this is also the solution to the original integer programming problem. If not all of the variables take integer values, further analysis is required, for example using branch-and-bound algorithms (e.g. Massart, Plastria and Kaufman, 1983). There is much symmetry in MP1, since the labelling of the classes is irrelevant; efficient algorithms prevent enumeration of equivalent solutions by imposing further constraints on class membership (e.g. Klein and Aronson, 1991).

The optimal facility location problem MP2 has attracted much attention. One general approach has involved Lagrangian relaxation of constraints (3.5) and (3.6), allowing bounds on optimal solutions to be derived and indicating if an optimal solution has been obtained (Daskin, 1995, Chapter 6). Another approach considers the dual of the problem in which condition (3.7) has been relaxed (Erlenkotter, 1978; Hanjoul and Peeters, 1985). As an illustration, the partition of the Abernethy Forest 1974 data into five classes which minimizes the sum of star distances $P(H4, \Sigma)$ was found to differ from the partition shown in Fig. 3.2 only in the assignment of sample number 32, the lowest member of class $B$, which was assigned to class $D$. The medoids of the five classes are shown circled in Fig. 3.2; some of the medoids are not located close to the centres of their classes in Fig. 3.2, but it should be remembered that this Figure provides a two-dimensional approximation to the nine-dimensional data on which the analysis was carried out.

Problem MP3 contains an exponential number of variables $(v_r)$, and would appear at first sight not to be computationally feasible for other than small values of $n$. However, a continuous relaxation of the integer programming problem can be tackled using column generation methodology (Gilmore and Gomory, 1961), in which the entering column in a revised simplex algorithm is selected from the solution of an auxiliary problem, further analysis being required if the relaxation fails to provide an integer solution. Problems which have been solved using this algorithm include the optimal facility location problem (Garfinkel, Neebe and Rao, 1974).

Fuller discussions of these and other mathematical programming algorithms are presented by Hansen, Jaumard and Sanlaville (1994) and Hansen and Jaumard (1997).

## 3.4 Other partitioning algorithms

This section presents a miscellaneous collection of other algorithms for obtaining partitions of a set of objects.

### 3.4.1 Minimum spanning trees and maximum-split partitions

Given $n$ vertices in a graph, a tree spanning these vertices is a set of edges such that the vertices belong to a connected graph which contains no cycles. When each of the $n(n-1)/2$ edges is assigned a length, a minimum spanning tree (MST) is a (not necessarily unique) spanning tree for which the sum of the edge lengths is smallest. In the classification context, each object in the data set is identified with a vertex of the graph, and the length of the edge joining the $i$th and $j$th vertices is set equal to $d_{ij}$, the dissimilarity between the $i$th and $j$th objects $(i, j = 1, ..., n)$.

The MST has been found useful in a wide variety of problems and several efficient algorithms for constructing it have been proposed; a historical account of the problem is presented by Graham and Hell (1985). An algorithm with $O(n^2)$ time complexity proposed by Kruskal (1956) can be described as follows:

> Repeatedly carry out the following step until the vertices all belong to a single component: amongst the edges not yet included in the graph, select for inclusion in the graph the shortest edge which does not form any cycles with edges already in the graph.

For example, the set of seven objects whose pairwise dissimilarities are given in Table 3.2 have the (unique) MST shown in Fig. 3.7, Kruskal's algorithm inserting the edges into the graph in the order (4,5), (3,6), (4,6), (6,7), (2,3), (1,4).

Removing from the MST the $(c-1)$ longest edges provides a maximum-split partition into $c$ classes, i.e. a partition which maximizes criterion $P(I1, min)$ (Zahn, 1971; Delattre and Hansen, 1980). The resulting classes are also single link classes (Florek et al., 1951): two objects, $i$ and $j$, are defined to belong to the same single link class at level $h$ if there exists a chain of $(m-1)$ intermediate objects, $i_1, i_2, ..., i_{m-1}$, linking them such that

$$d_{i_k, i_{k+1}} \leq h \text{ for } k = 0, 1, ..., m-1 \ (1 \leq m \leq n-1),$$

where $i_0 \equiv i$ and $i_m \equiv j$. The value of $h$ controls the scale of the investigation, with different values of $h$ providing partitions comprising possibly different numbers of single link classes. By allowing $h$ to range over the values in the dissimilarity matrix, a hierarchically-nested set of partitions is obtained; further details of the single link method are presented in Section 4.2.2. Single link classes may possess little within-class homogeneity, as is illustrated

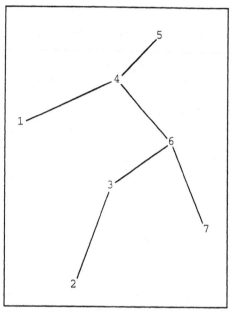

Figure 3.7 *The minimum spanning tree of the set of objects whose dissimilarity matrix is given in Table 3.2.*

by the class comprising objects {3 - 7} formed by the deletion from Fig. 3.7 of edges (2,3) and (1,4).

Fig. 3.8 superimposes the minimum spanning tree of the Abernethy Forest 1974 data on the first two principal components of that data set. There are several places where two points that are neighbours in the plot are not directly linked in the MST, providing an indication of the extent to which information about the nine-dimensional data set is lost in the two-dimensional approximation. The four longest edges in the MST are shown by dashed lines. Removing these edges provides a maximum split partition into five classes; these are the same five classes obtained from the minimization of the sum of squares $P(H1, \Sigma)$, as may be seen by comparing Figs. 3.2 and 3.8.

### 3.4.2 Hybrid classification and assignment algorithms

Algorithms have been proposed which select and classify a subset ($\omega$, say) of the set of objects, and then assign each of the objects not

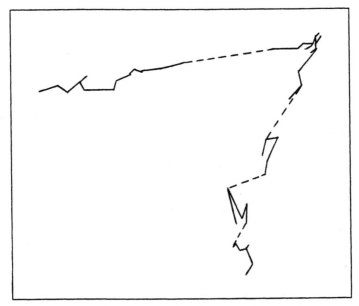

Figure 3.8 *The minimum spanning tree of the Abernethy Forest 1974 data, with the four longest edges shown by dashed lines.*

belonging to $\omega$ to the class in the partition to which it is 'closest'. There have been two main motivations for such work. First, it provides an approach which is relevant for the analysis of large data sets. Secondly, it can be helpful in classifying objects for which the 'nuclei' of classes can be identified but the boundaries between classes are not clear-cut: for data represented by points in some space, nuclei of classes correspond to locally dense regions of points surrounded by less dense regions.

The results can depend markedly on the specification of the subset $\omega$: ideally, it should contain representatives of all classes that are present (but still require to be identified) in the complete data set, although it may be unrealistic to expect objects from small classes to be included in $\omega$. In some problems, the selection can be guided by external information, such as the spatial location of objects in remote sensing studies (e.g. Bryant, 1979). In the absence of such external information, one can construct $\omega$ by sequentially adding to it objects which have many close neighbours. In carrying out this process, it seems preferable to consider as neighbours only those objects which are not yet themselves members of $\omega$, otherwise $\omega$ can be dominated by objects belonging to the denser nuclei. The members of $\omega$ can be sequentially classified as they are selected (Wishart, 1969a) or classification of them can be postponed until the entire subset $\omega$ has been selected (Gordon, 1986a).

One method of implementing these ideas for constructing the subset $\omega$ involves examining the set of objects that have not yet been selected and, for each object, evaluating its average dissimilarity with its $k$ nearest neighbours for several small values of $k$. Either by specifying a single value of $k$, or by examining a plot of each object's average dissimilarity over a range of small values of $k$, the next object for inclusion in $\omega$ can be identified; the average dissimilarities are then recalculated. For large data sets, distributed array processors (e.g. Gostick, 1979) provide an efficient method of evaluating average dissimilarities and constructing the subset $\omega$.

After the subset of objects has been classified using an appropriate clustering criterion, each of the other objects is assigned to the class to which it is 'closest'; adaptive measures of distance, as described in Section 3.2.3, seem particularly appropriate for the assignment. Depending on the size of the data set, it may then be relevant to apply an iterative relocation algorithm to this partition of the complete data set.

As an illustration of this methodology, an analysis is presented of the data plotted in Fig. 3.9. These data comprise 100 observations from each of three bivariate normal distributions. The centres of the distributions are located at the midpoints of the sides of an equilateral triangle whose sides are of length 10. For each of the three distributions, the major axis of its variance-covariance matrix lies along the side of the triangle and has length 4, with the minor axis having length 1. There is a fair amount of overlap between the samples from the different distributions.

A subset of 75 objects was selected from dense regions of the plane by sequentially identifying objects with minimum average distance to their fifth nearest neighbour (amongst objects that had not yet been selected). The 75 selected objects are plotted in Fig. 3.10. The visually-evident partition of these objects into three classes is confirmed by several different classification criteria, but it should be noted that one of the objects in the right-hand class (shown circled in Fig. 3.10) was actually generated from the distribution that provided the objects plotted at the bottom of

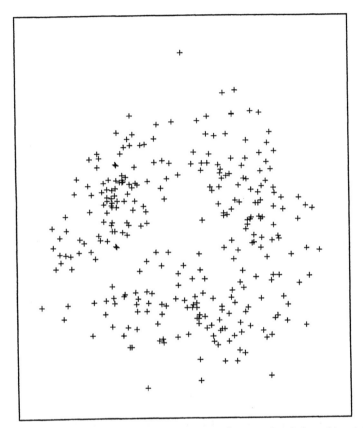

Figure 3.9 *A plot of samples of size 100 from each of three bivariate normal distributions.*

the Figure. The sample variance-covariance matrices of these three classes were evaluated and all 300 objects were assigned to the class whose Mahalanobis distance to them was smallest. The resulting partition is shown in Fig. 3.11, in which the convex hull of each class has been drawn and the 19 objects which are assigned to classes other than the one specified by their parent distribution are shown circled.

### 3.4.3 Identification of individual classes

The work in the earlier sections of this chapter has involved

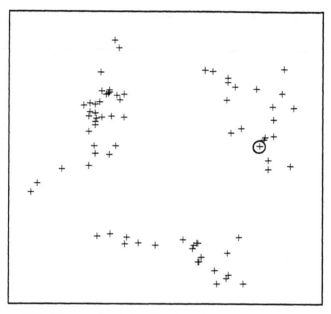

Figure 3.10 *The sample of size 75 selected from the data shown in Fig. 3.9.*

- defining criteria to be satisfied by an optimal partition of a set of objects into a stated number of classes

- describing algorithms which seek optimal partitions.

This section considers the same agenda for individual classes of objects. The set of classes identified may provide a partition of the complete set of objects; alternatively, some objects may belong to none of the classes, or to more than one class.

Two 'ideal' classes are defined as follows: $C$ is a 'comprehensive type' (McQuitty, 1963, 1967) or 'ball' (Jardine, 1969) cluster if for each object $i \in C$,

$$\max_j\{d_{ij} \mid j \in C\} < \min_k\{d_{ik} \mid k \notin C\}; \qquad (3.13)$$

$C$ is an $L^*$-cluster (van Rijsbergen, 1970) if

$$\max\{d_{ij} \mid i,j \in C\} < \min\{d_{kl} \mid k \in C, l \notin C\}. \qquad (3.14)$$

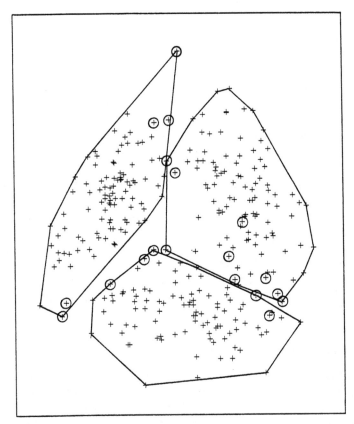

Figure 3.11 *A partition of the data shown in Fig. 3.9; misclassified points are circled.*

In terms of the heterogeneity and isolation criteria defined in Section 3.1, $C$ is an $L^*$-cluster if its diameter $H3(C)$ is less than its split $I1(C)$.

$L^*$-clusters are also ball clusters, and both are contained in the set of single link classes. Single link classes of size two automatically satisfy conditions (3.13) and (3.14), but these conditions become increasingly stringent as the size of the class increases and few data sets contain large ball clusters or $L^*$-clusters. The only $L^*$-clusters contained in the data set whose dissimilarity matrix is given in Table 3.2 are {4, 5} and {3, 6}. However, in the Abernethy

Forest 1974 data set, there are six ball clusters of size greater than two, including classes $C$ and $D$ (containing, respectively, seven and five objects: see Fig. 3.2), all of these ball clusters also being $L^*$-clusters; this indicates a very clear cluster structure in this part of the data set.

Conditions (3.13) and (3.14) provide a complete definition of the classes to be found. Some alternative approaches require the specification of parameters. For example, given a suitable measure of the heterogeneity (resp., isolation) of a class, such as those defined in Section 3.1, one can seek *(i)* classes whose measure does not exceed (resp., does not fall below) a specified threshold value, $T$, or *(ii)* classes of size $k$ which are optimal over the set of classes of size $k$ (e.g. Hubert, 1974a; Hansen, Jaumard and Mladenovic, 1995). This requires a method of identifying appropriate values of $T$ or $k$, e.g. values for which small changes would lead to a marked deterioration in the solution. The classes identified would usually be removed from the data set and the analysis be repeated on the remaining objects; alternatively, for some criteria, classes may always be identified from the complete data set, allowing the possibility of obtaining overlapping classes (e.g. Lefkovitch, 1980).

## 3.5 How many clusters?

Earlier sections of this chapter have described how one can obtain a partition of a set of $n$ objects into a specified number, $c$, of classes or clusters. It has been assumed that appropriate values of $c$ are known at the start of each investigation, but this is rarely the case. This section discusses ways in which researchers have attempted to answer the question "how many clusters are there in the data set?"

The most common approach is to obtain 'optimal' partitions into $c$ classes for a range of values of $c$. Usually, a complete set of hierarchically-nested partitions is obtained using an agglomerative algorithm. Such algorithms are described in Chapter 4; briefly, the initial partition is into $n$ singleton clusters and each subsequent partition is obtained by amalgamating a pair of 'closest' clusters, where different definitions of what is meant by 'closest' correspond to different clustering criteria. The set of partitions into $c$ clusters $(1 \leq c \leq n)$ is then examined in an attempt to determine the most appropriate value of $c$. An informal assessment involves studying graphical representations of the data onto which class boundaries indicated by clustering criteria have been superimposed, as is illustrated in Figs. 3.2 and 3.3; this allows one to decide if the objects fall naturally into a certain number of classes, while reducing the subjectivity associated with unaided visual assessment of plots.

Some more formal methods of deciding upon appropriate values of $c$ have been collectively referred to as 'stopping rules', as one could envisage stopping the amalgamation process at the selected value of $c$. It is convenient to categorize stopping rules as either *global* or *local*. Global rules evaluate a measure, $G(c)$, of the goodness of the partition into $c$ clusters, usually based on the within-cluster and between-cluster variability, and identify the value of $c$ for which $G(c)$ is optimal. A disadvantage of many of these rules is that there is no natural definition of $G(1)$, hence they can provide no guidance on whether or not the data *should* be partitioned. Local rules involve examining whether or not a pair of clusters should be amalgamated (or a single cluster should be subdivided). Unlike global rules, they are thus based on only part of the data (except when the comparison is being made between $c = 1$ and $c = 2$) and can assess only hierarchically-nested partitions. A disadvantage of local rules is that they generally need the specification of a threshold value or significance level, the optimal value of which will depend on the (unknown) properties of the data set that is under investigation.

Many different stopping rules have been proposed, often with cursory investigations of their properties and only limited comparisons of them with previously-published rules. The most detailed comparative study that has been carried out appears to be that undertaken by Milligan and Cooper (1985). These authors investigated the extent to which 30 published stopping rules were able to detect the correct number of clusters in a series of simulated data sets, constructed to contain fairly clear-cut cluster structure. The clusters in their study comprised mildly truncated data from multivariate normal distributions, and one would not expect their ranking of the set of stopping rules to be reproduced exactly if a different cluster-generating strategy were adopted. Further, one would expect different stopping rules to be well-suited for use in analysing clusters provided by different clustering criteria. Nevertheless, Milligan and Cooper's (1985) study is valuable for identifying stopping rules which performed poorly, and which cannot therefore be recommended for further use.

The five rules (three global and two local) which performed best in Milligan and Cooper's (1985) study are described below.

$G1$: Caliński and Harabasz's (1974) index for assessing a partition of a set of objects described by quantitative variables is defined by

$$G1(c) \equiv [B/(c-1)]/[W/(n-c)],$$

where $W$ and $B$ denote, respectively, the total within-cluster sum of squared distances about the centroids, and the total between-cluster sum of squared distances.

$G2$: Variants of Goodman and Kruskal's (1954) $\gamma$ have been widely used in classification studies. In the present context, comparisons are made between all within-cluster dissimilarities and all between-cluster dissimilarities. A comparison is deemed to be concordant (resp., discordant) if a within-cluster dissimilarity is strictly less (resp., strictly greater) than a between-cluster dissimilarity; equalities between members of the two sets of dissimilarities are disregarded in the definition of the index.

$$G2(c) \equiv (S_+ - S_-)/(S_+ + S_-),$$

where $S_+$ (resp., $S_-$) denotes the number of concordant (resp., discordant) comparisons.

$G3$: This index standardizes the sum, $D(c)$, of all within-cluster dissimilarities in a partition of the objects into $c$ clusters. If the partition has a total of $r$ such dissimilarities, $D_{min}$ (resp., $D_{max}$) is defined to be the sum of the $r$ smallest (resp., largest) dissimilarities, and

$$G3(c) \equiv (D(c) - D_{min})/(D_{max} - D_{min}).$$

The value of $c$ which maximizes $G1(c)$ or $G2(c)$, or minimizes $G3(c)$, is regarded as specifying the number of clusters in the data set – with the restriction that only smaller values of $c$ are investigated, as some indices can display distracting behaviour for larger values of $c$ (Milligan and Cooper, 1985).

$L1$: Duda and Hart (1973, Section 6.12) proposed a rule for deciding whether or not a cluster should be divided into two sub-clusters, based on comparing its within-cluster sum of squared distances $(W_1)$ with the sum of within-cluster sum of squared distances when the cluster was optimally divided into two $(W_2)$. If the cluster contains $m$ objects described by $p$ quantitative variables, the hypothesis that the cluster is homogeneous (and hence should not be subdivided) is rejected if

$$W_2/W_1 < 1 - 2/(\pi p) - z[2(1 - 8/(\pi^2 p))/(mp)]^{1/2},$$

where $z$ is a standard normal deviate specifying the significance level of the test.

$L2$: Beale (1969) proposed a test for deciding if a cluster should be subdivided, involving the comparison of

$$F \equiv \left(\frac{W_1 - W_2}{W_2}\right) \bigg/ \left(\left(\frac{m-1}{m-2}\right) 2^{2/p} - 1\right)$$

(where $W_1$, $W_2$, $m$ and $p$ are as defined in $L1$) with an $F_{p,(m-2)p}$ distribution and the rejection of the hypothesis of a single cluster for significantly large values of $F$.

In both of these local stopping rules, amalgamation has usually proceeded until the hypothesis of a single cluster is first rejected, although one could examine the sets of $z$ and $F$ values resulting from the local stopping rules in order to identify particularly significant results (in effect, converting these tests into global stopping rules). Clearly, significance levels should not be interpreted strictly, because a set of related investigations is being carried out.

It is advisable not to depend on a single stopping rule, but to synthesize the results of several. All five tests have been used to assess the number of clusters present in the Abernethy Forest 1974 data. A complete hierarchical classification of these data was obtained using an agglomerative algorithm which at each stage amalgamates a pair of clusters which leads to minimum increase in the total within-cluster sum of squared distances (Ward (1963), see Chapter 4), because the partition that is obtained when the hierarchy is sectioned may be regarded as an approximation to the optimal sum of squares partition into the relevant number of clusters. Fig. 3.12 plots the values of the three global indices for values of $c$ between 2 and 15, and the values of $z$ and $F$ obtained from the two local stopping rules for each value of $c$ between 1 and 15;

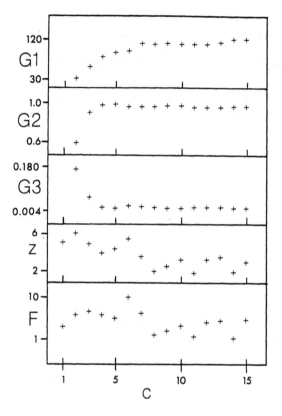

Figure 3.12 *Application of three global stopping rules and two local stopping rules to the Abernethy Forest 1974 data: the values of various criteria are plotted against the number of clusters, c, in the partition.*

although the critical values of an $F$ distribution depend on its degrees of freedom, the variation is not large for the values of $p$ and $m$ considered for these data.

The results shown in Fig. 3.12 are somewhat contradictory. $G1(c)$ has local maxima at $c = 9$ and $c = 7$, but increases fairly steadily with $c$, increasing further for values of $c$ above 15. $G2(c)$ has maximum value when $c = 5$, this value remaining the largest for all $c < 32$. $G3(c)$ has a local minimum at $c = 5$, but $G3(5) > G3(15)$ and the downward trend continues for higher values of $c$. The two local stopping rules illustrate the problem of specifying a relevant threshold or critical value. If one examines the entire set of values shown in Fig. 3.12, the high values at $c = 6$ (when $z = 5.57$ and $F = 9.92$, corresponding to extremely small levels of significance) suggest that one might consider there to be seven clusters. Overall, one might conclude that there are five, or possibly seven, clusters in the data set. The five clusters are shown in Fig. 3.2, the seven clusters being obtained from them by subdividing clusters $A$ and $B$. A similar investigation of the 90 objects from the combined Abernethy Forest data suggested that they could be partitioned into eight clusters.

Methodology is described in Chapter 7 for more formal investigations of aspects of classifications.

## 3.6 Links with statistical models

Automated methods of classification were perceived by some of their early users as having the advantage of providing analyses of data which did not require the specification of a (possibly inappropriate) statistical model. However, classification criteria *do* involve assumptions about the type of class structure that is present in data, and can distort the results of an analysis towards finding certain types of class; in effect, there is a model for the data, even though it is not explicitly stated.

For example, investigators initially used the sum of squares criterion because it appeared to provide an intuitively reasonable definition of compact clusters. It has been observed empirically that the classes resulting from use of this criterion have a tendency to contain roughly equal numbers of objects and to occupy hyperspherical regions in space, as illustrated in Fig. 3.5. More formal modelling explains why one would expect such results to occur.

Thus, Scott and Symons (1971) consider a multivariate normal components model, in which the $r$th population is assumed to have mean vector $\mu_r$ and covariance matrix $\Sigma_r$ and to contribute $n_r$ objects to the data set $(r = 1, ..., c; n = \Sigma_{r=1}^{c} n_r)$. The identity of the $n_r$ objects belonging to the $r$th population and the value of $n_r$ are assumed to be unknown. A vector of classification parameters $\gamma \equiv (\gamma_1, ..., \gamma_n)$ is defined by $\gamma_i = r$ if the $i$th object belongs to the $r$th population $(i = 1, ..., n; r = 1, ..., c)$.

In this 'fixed partition' model, the main interest is in identifying the partition specified by $\gamma$, and the estimates of the set of parameters

$$\theta \equiv (\gamma, \mu_1, ..., \mu_c, \Sigma_1, ..., \Sigma_c)$$

which maximize (3.15) are called 'classification maximum likelihood' estimates. Such estimates can be found under a variety of assumptions about the parameters $\{\Sigma_r (r = 1, ..., c)\}$.

Given data $\{x_i (i = 1, ..., n)\}$, the log likelihood function is

$$l(\theta) = k - \Sigma_{r=1}^{c}[\Sigma_{i \in E_r}\{(x_i - \mu_r)'\Sigma_r^{-1}(x_i - \mu_r)\} + n_r \text{logdet}(\Sigma_r)]/2,$$
(3.15)

where $k$ is a constant and $E_r$ denotes the set of labels $i$ for which $\gamma_i = r$.

For a given partition of the objects into $c$ classes, $l(\theta)$ is maximized by substituting the maximum likelihood estimates $\hat{\mu}_r$ and $\hat{\Sigma}_r$ of $\mu_r$ and $\Sigma_r$; irrespective of assumptions made about $\{\Sigma_r (r = 1, .., c)\}$,

$$\hat{\mu}_r = \bar{x}_r \equiv \Sigma_{i \in E_r} x_i / n_r.$$
(3.16)

By substituting from (3.16) into (3.15) and using the fact that a scalar is equal to its trace, the maximum likelihood estimate $\hat{\gamma}$ of $\gamma$ is obtained from the minimization of

$$\Sigma_{r=1}^{c}[\text{trace}(W_r\Sigma_r^{-1}) + n_r \text{logdet}(\Sigma_r)],$$
(3.17)

where

$$W_r \equiv \Sigma_{i \in E_r}(x_i - \bar{x}_r)(x_i - \bar{x}_r)',$$
(3.18)

the cross-product matrix for the class of objects $E_r$ $(r = 1, ..., c)$.

If $\Sigma_r = \sigma^2 I$ $(r = 1, ..., c)$, the maximum likelihood estimate of $\gamma$ is obtained by minimizing the trace of the matrix $W \equiv \Sigma_{r=1}^{c} W_r$, i.e. by minimizing the total within-class sum of squares about the class centroids. In other words, use of the sum of squares criterion $P(H1, \Sigma)$ defined in Section 3.1 is equivalent to maximum likelihood estimation under a spherical normal components model.

Different assumptions about $\{\Sigma_r(r = 1, ..., c)\}$ lead to different criteria to be optimized. If $\{\Sigma_r(r = 1, ..., c)\}$ are assumed to be all equal to $\Sigma$, the maximum likelihood estimate of $\Sigma$ is $\hat{\Sigma} = W/n$; when this estimate is substituted into (3.17), maximum likelihood estimation of $\gamma$ is seen to involve the minimization of det $W$, a criterion proposed by Friedman and Rubin (1967). Banfield and Raftery (1993) express $\Sigma_r$ in terms of its eigenvalue decomposition:

$$\Sigma_r \equiv V_r \Lambda_r V_r',$$

where $\Lambda_r$ is a diagonal matrix containing the eigenvalues of $\Sigma_r$ and $V_r$ is the matrix of eigenvectors. If $\lambda_r$ denotes the largest eigenvalue of $\Lambda_r$ and the diagonal matrix $A_r$ is defined by

$$A_r \equiv \lambda_r^{-1} \Lambda_r,$$

$V_r$, $A_r$ and $\lambda_r$ can be regarded as providing measures of, respectively, the orientation and shape of the $r$th class and its volume in $p$-dimensional space. Banfield and Raftery (1993) investigate the classification criteria that result when selected combinations of these measures are varied. Marriott (1982) studies the properties of several classification criteria by examining the effect of adding an extra object to the data set.

In the classification maximum likelihood approach, interest is concentrated on the assignment of objects to classes that is specified by $\gamma$. If one were interested in estimating parameters of the multivariate normal populations, it would not be advisable to estimate the parameters of each population from the information provided by a single class, because the estimators need be neither unbiassed nor consistent (Marriott, 1975; Bryant and Williamson, 1978). Symons (1981), Windham (1987) and Bryant (1991) describe methodology which combines assignment of objects and estimation of parameters.

The implications of specifying prior distributions for parameters of models in a Bayesian context have also been investigated (e.g. Scott and Symons, 1971; Binder, 1978; Symons, 1981). For example, with $\Sigma_r = \Sigma$ $(r = 1, ..., c)$, Scott and Symons (1971) assume a prior density for the parameters proportional to

$$\pi(\gamma)[\det(\Sigma)]^{-(p+1)/2}$$

over the region of interest, and show that the assignment of objects that has highest posterior probability corresponds to the maximum likelihood estimate if

$$\pi(\gamma) \propto \prod_{n_k > 0} n_k^{p/2}.$$

This prior probability heavily weights classes of equal size, in conformity with observed results.

Such links between statistical models and classification criteria illustrate the fact that classification criteria are not 'model-free', and enable investigators to make more informed choices of relevant methods of analysis for their data. A detailed survey of the topic is presented by Bock (1996).

# CHAPTER 4

# Hierarchical classifications

## 4.1 Definitions and representations

The previous chapter described ways of defining and obtaining a partition of a set of objects. On occasion, it is relevant to obtain fuller information about a set of objects by obtaining a hierarchical-ly-nested set of such partitions; for example, in taxonomy an object could belong successively to a species, a genus, a family, an order, etc., and the relationships between different classes of objects are of interest.

Two constructs that are relevant in this context are an *n-tree* and a *valued tree*. An *n*-tree (Bobisud and Bobisud, 1972; McMorris, Meronk and Neumann, 1983) is a set $T$ of subsets of the set of objects $\Omega$ satisfying the following conditions:

$(i)$ $\Omega \in T$; $(ii)$ $\emptyset \notin T$; $(iii)$ $\{i\} \in T$ for all $i \in \Omega$;
$(iv)$ if $A, B \in T$, then $A \cap B \in \{\emptyset, A, B\}$.

This is illustrated by the tree diagram shown in Fig. 4.1, in which:

- each of the nine objects is located at a terminal node (or 'leaf') of the tree

- each of the internal nodes, labelled $A - G$, represents the class of objects subtended by the node; the node $G$ is called the *root* of the tree.

Thus, in addition to the complete set of objects $\Omega \equiv \{1, 2, ..., 9\}$ and the singleton subsets $\{i\}$ $(i = 1, ..., 9)$, which belong to all *n*-trees, the tree also contains the subsets $\{8, 4\}, \{2, 6, 3\}, \{1, 5\}$, $\{8, 4, 9\}, \{1, 5, 7\}$ and $\{8, 4, 9, 2, 6, 3, 1\}$, corresponding to the internal nodes $A - F$, respectively.

Mathematicians will recognise an *n*-tree as an unordered rooted tree with labelled leaves, in which none of the internal nodes, except possibly the root, has degree (i.e. number of incident edges) two. If the root is of degree two and all other internal nodes are of degree

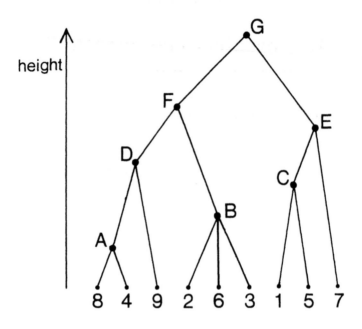

Figure 4.1 *A rooted tree.*

three, the tree is called *binary* or *fully resolved*. A binary tree has $n - 1$ internal nodes, and each of its non-singleton classes has only two offspring classes. The tree shown in Fig. 4.1 is not binary, since subset $B$ has three offspring (singleton) subsets.

An $n$-tree specifies only the subsets belonging to a hierarchical classification. There have been several characterizations (e.g. Hartigan, 1967; Jardine, Jardine and Sibson, 1967; Johnson, 1967) of a valued tree (Boorman and Olivier, 1973), in which a *height* $h$ is associated with each of these subsets. Nested subsets, such as $A$ and $D$ in Fig. 4.1, satisfy the condition:

$$h(A) \leq h(D) \Leftrightarrow A \subseteq D.$$

For each pair of objects $(i, j)$, $h_{ij}$ $(= h_{ji})$ is defined to be the height of the smallest subset containing both the $i$th and $j$th objects. The value of $h_{ij}$ measures the difference between the $i$th and $j$th objects in the classification, with small values of $h_{ij}$ indicating that the corresponding objects are perceived as similar to one another.

The set of values $\{h_{ij}\ (i, j \in \Omega)\}$ satisfies the ultrametric inequality:

$$h_{ij} \leq \max(h_{ik}, h_{jk}) \text{ for all } i, j, k \in \Omega. \tag{4.1}$$

An equivalent condition to (4.1) is that every triple of objects $(i, j, k)$ possesses the property that the two largest values in the set $\{h_{ij}, h_{ik}, h_{jk}\}$ are equal.

Although hierarchical classifications of sets of objects generally specify ultrametric distances $(h_{ij})$, investigators rarely make use of the values themselves, largely because there is often interest in comparing different classifications of the same set of objects and ultrametric distances resulting from use of different classification criteria are not commensurate. Instead, attention is more commonly concentrated on *ranked trees*, in which only the ordering of the heights of the internal nodes is considered. The only ordering of subsets involved in $n$-trees is that required by set inclusion, and $n$-trees are sometimes alternatively referred to as non-ranked trees (Murtagh, 1984).

There are several ways in which valued trees can be represented graphically. The most common way is as tree diagrams like the one shown in Fig. 4.1, which are usually referred to as dendrograms, although it should be noted that the word 'dendrogram' is sometimes taken to refer also to valued trees and not just to their pictorial representations. There are several different formats in which dendrograms have been presented, as illustrated in Fig. 4.2. The most common formats are those shown in ($a$) and ($b$) in Fig. 4.2, or versions of them rotated through 90 degrees or 180 degrees (some investigators prefer the root of a tree to lie below its leaves). Fig. 4.2($c$) shows a right-justified dendrogram, in which each class is labelled by its rightmost member: this format highlights the fact that a hierarchical classification can be defined by specifying for each object

- the class which it joins when it ceases to be the rightmost member of its class, and

- the height at which this occurs.

Fig. 4.2($d$) shows a *stick tree* (Hartigan, 1975, Chapter 10; Kruskal and Landwehr, 1983), with the property that the membership of each class is indicated by the extent of a horizontal line.

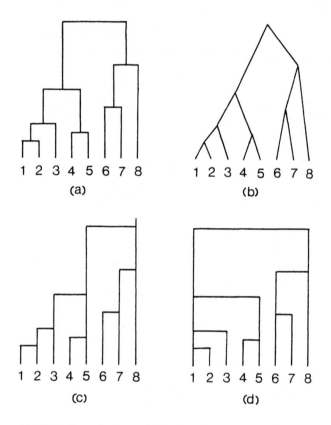

Figure 4.2 *Four formats for representing the same hierarchical classification of the set of objects* $\Omega \equiv \{1, 2, ..., 8\}$.

Even after the dendrogram format has been selected, the representation of a hierarchical classification still contains much indeterminacy, since one can permute the left-right ordering of the edges leading down from each internal node: thus, there are $2^{n-1}$ equivalent representations of a binary valued tree. The ease of interpretation of results can be increased by suitable permutation of the edges subtended by each internal node. For example, the ordering of the leaves of the dendrogram can be chosen to maximize a measure of their rank correlation with an externally-provided seriation of the objects. The seriation could be specified to be that

Table 4.1 *An icicle plot of the hierarchical classification represented in Fig. 4.2.*

| 1 | = | 2 | = | 3 | = | 4 | = | 5 | = | 6 | = | 7 | = | 8 |
|---|---|---|---|---|---|---|---|---|---|---|---|---|---|---|
| 1 | = | 2 | = | 3 | = | 4 | = | 5 | | 6 | = | 7 | = | 8 |
| 1 | = | 2 | = | 3 | = | 4 | = | 5 | | 6 | = | 7 | | |
| 1 | = | 2 | = | 3 | | 4 | = | 5 | | 6 | = | 7 | | |
| 1 | = | 2 | = | 3 | | 4 | = | 5 | | | | | | |
| 1 | = | 2 | | | | 4 | = | 5 | | | | | | |
| 1 | = | 2 | | | | | | | | | | | | |

ordering of the rows and columns of the dissimilarity matrix which satisfies as closely as possible the property that values within the matrix are monotonically non-decreasing as one moves away from the diagonal within each row and within each column (Robinson, 1951; Gale, Halperin and Costanzo, 1984). In this book, a matrix with this property is referred to as a Robinson matrix, but it should be noted that some authors have referred to such a matrix as 'anti-Robinson', reserving the adjective 'Robinson' for a (similarity) matrix whose values are monotonically non-increasing as one moves away from the diagonal within each row and within each column.

In a set of alternative representations, the extent of a class is shown by a solid area. Examples of such representations are skyline plots (Wirth, Estabrook and Rogers, 1966), icicle plots (Kruskal and Landwehr, 1983) and banners (Rousseeuw, 1986), whose main difference from one another is in the orientation of the display. The hierarchical classification depicted by dendrograms in Fig. 4.2 is summarized in Table 4.1 by an icicle plot, in which singleton classes are suppressed and objects in the same class are linked by '='; for example, the longest icicle (1 = 2) indicates that class {1,2} is the first to be formed. Particularly for large data sets, such plots can assist an investigator in the rapid identification of the membership of each class.

The information contained in hierarchical classifications can also be superimposed onto graphical representations, in which the objects are represented by a set of points in a two-dimensional space and the objects belonging to each class are surrounded by a closed curve (Hubac, 1964; Moss, 1967; Rohlf, 1970). Fig. 4.3 shows one

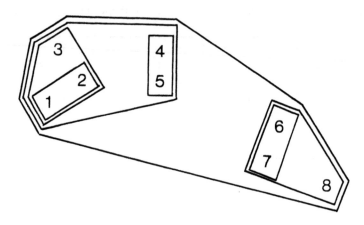

Figure 4.3 *A superimposition of a hierarchical classification of eight objects onto a graphical representation.*

such plot for the hierarchical classification represented by dendrograms in Fig. 4.2. The positioning of the points in the graphical representation can be obtained using methodology described in Chapter 6, and can provide supplementary information about the resemblances of the objects. Plots like Fig. 4.3 are particularly relevant for representing $n$-trees, since only the subset information is retained. Two variants of Fig. 4.3 can be used to represent ranked trees and valued trees, respectively. First, for each of the partitions comprising the hierarchical classification, a separate bounding curve can be drawn round each of its constituent classes; thus, classes which remain unaltered while others are changing will be surrounded by a collection of bounding curves. Secondly, the height of each class can be indicated by associating a 'contour' value with each of the bounding curves shown in Fig. 4.3.

Generalizations of dendrograms have been proposed which provide information about two different characteristics of the classes in a tree diagram, such as measures of their heterogeneity and isolation as defined in Section 3.1. The value of the second characteristic can be represented by the horizontal distance between the leaves (McCammon, 1968) or by the lengths of the horizontal lines in Fig. 4.2(a). For example, Fig. 4.4 shows an *espalier*, in which the split of a class is recorded on the vertical axis and the diameter

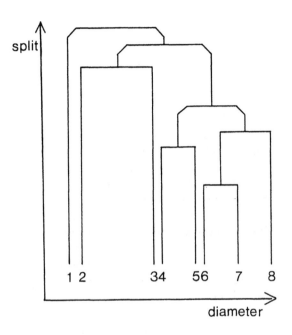

Figure 4.4 *An espalier depicting the split and diameter of classes in a hierarchical classification.*

of a class is shown by the length of the horizontal line above it. Espaliers can be constructed for a given ordering of the leaves, or the ordering of offspring classes can be permuted so as to minimize the width of the espalier; relevant methodology and algorithms are described by Hansen, Jaumard and Simeone (1996).

## 4.2 Algorithms for obtaining hierarchical classifications

### 4.2.1 Direct optimization algorithms

The problem of obtaining a hierarchical classification of a set of objects can be formulated in terms of transforming a matrix of pairwise dissimilarities $(d_{ij})$ into a matrix of ultrametric distances $(h_{ij})$ satisfying condition (4.1). It is natural to define a measure of the discordance between $(d_{ij})$ and $(h_{ij})$, and to seek ultrametric distances $\mathbf{h} \equiv (h_{ij})$ which minimize the stated measure of discordance. However, Křivánek and Morávek (1986) and Křivánek (1986) showed that this problem is NP-hard when the measure of discordance is either

$$L_1(\mathbf{h}) \equiv \Sigma_{1 \leq j < i \leq n} \mid d_{ij} - h_{ij} \mid \qquad (4.2)$$

or

$$L_2(\mathbf{h}) \equiv \Sigma_{1 \leq j < i \leq n} (d_{ij} - h_{ij})^2. \qquad (4.3)$$

Optimal solutions can be obtained for small data sets using branch-and-bound and dynamic programming methodology (Chandon, Lemaire and Pouget, 1980; Hubert, Arabie and Meulman, 1997). For larger data sets, various approximating algorithms have been published. In particular, Carroll and Pruzansky (1975, 1980) and De Soete (1984a) augment a least squares measure of discordance by adding a penalty function which measures the extent to which $(h_{ij})$ fails to satisfy the ultrametric condition. Given that for $(h_{ij})$ to satisfy the ultrametric inequality it is necessary for the largest two of $\{h_{ij}, h_{ik}, h_{jk}\}$ to be equal for all $i, j, k \in \Omega$, the penalty function is defined by

$$P(\mathbf{h}) \equiv \Sigma_U (h_{ik} - h_{jk})^2, \qquad (4.4)$$

where

$$U \equiv \{(i, j, k) \mid h_{ij} \leq \min(h_{ik}, h_{jk})\}, \qquad (4.5)$$

and the overall function to be minimized is

$$F(\mathbf{h}, \alpha) \equiv L_2(\mathbf{h}) + \alpha P(\mathbf{h}). \qquad (4.6)$$

An iterative function minimization algorithm is used to obtain solutions for increasing values of $\alpha$, driving the value of the penalty function to zero.

An alternative algorithm for obtaining hierarchical classifications is based on iterative projection onto closed convex sets defined by sets of ultrametric inequalities (Hubert and Arabie, 1995).

This methodology is illustrated by application to the acoustic confusion data. De Soete (1984a) used the penalty function approach to seek a matrix of ultrametric distances which provides a least squares fit to a dissimilarity matrix for these data, his results being summarized in the dendrogram shown in Fig. 4.5. The two largest classes in this hierarchical classification correspond

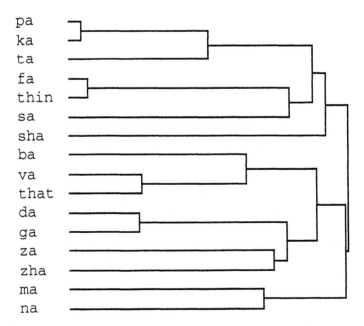

Figure 4.5 *Least squares ultrametric fit to the acoustic confusion data.*

to the unvoiced consonants $\{pa, ka, ta, fa, thin, sa, sha\}$ and the voiced consonants $\{ba, va, that, da, ga, za, zha, ma, na\}$. The stops $\{ta, pa, ka\}$ form a subset of the unvoiced consonants, and the nasals $\{ma, na\}$ clearly differ from the other members of the voiced consonants. Further discussions of these data are given by Shepard (1972), Arabie and Carroll (1980), Carroll and Pruzansky (1980) and De Soete (1984a), and a different analysis of them is described in Section 5.3.1.

Direct optimization algorithms have also been described for obtaining hierarchical classifications based on incomplete dissimilarity matrices (De Soete, 1984b) and matrices in which information is available only about the dissimilarities between objects belonging to two disjoint sets, $(d_{ij} \mid i \in \Omega_1; j \in \Omega_2; \Omega_1 \cap \Omega_2 = \emptyset)$ (De Soete et al., 1984); in the latter case, the relationships within classes containing objects from a single set, $\Omega_r$, cannot be uniquely described.

### 4.2.2 Agglomerative algorithms

Agglomerative algorithms start with each object in a separate class. At each step of the algorithm, the 'most similar' classes are amalgamated, where the definition of 'most similar' is provided by the clustering criterion being used. Amalgamation continues until all the objects belong to a single class. The most widely used algorithms amalgamate a single pair of classes at each step; unless subsequent additions to this class can occur at the same height, this imposes a binary tree on the data, which may be inappropriate (consider class $B$ in Fig. 4.1). Further, the 'most similar' pair of classes might not be uniquely defined and the choice made when there are ties can markedly influence the results (Hart, 1983; Morgan and Ray, 1995); there remains a need for computer software that warns investigators of the existence of multiple solutions.

Agglomerative algorithms are 'stepwise optimal': at each step, the amalgamation chosen is the best (in terms of the specified clustering criterion) that can be made at that time. However, for most clustering criteria, it is not possible to state that the complete hierarchical classification satisfies any optimality criterion.

The most commonly used agglomerative algorithms can be described in terms of a recurrence relation in which the dissimilarity between a newly-merged class, $C_i \cup C_j$, and any other class, $C_k$, is defined by:

$$
\begin{aligned}
d(C_i \cup C_j, C_k) = {} & \alpha_i d(C_i, C_k) + \alpha_j d(C_j, C_k) + \beta d(C_i, C_j) \\
& + \gamma \mid d(C_i, C_k) - d(C_j, C_k) \mid + \delta_i h(C_i) \\
& + \delta_j h(C_j) + \epsilon h(C_k),
\end{aligned} \tag{4.7}
$$

where $h(C_i)$ is the height in the valued tree of class $C_i$, and

$$
\theta \equiv \{\alpha_i, \alpha_j, \beta, \gamma, \delta_i, \delta_j, \epsilon\}
$$

is a set of parameters whose values specify the clustering criterion. A recurrence relation containing the terms with coefficients $\alpha_i, \alpha_j, \beta$ and $\gamma$ was proposed by Lance and Williams (1966b, 1967) and the generalization which includes the other three terms was suggested by Jambu (1978).

Initially, $C_i = \{i\}, C_j = \{j\}$ and $d(C_i, C_j) = d_{ij}(i, j = 1, ..., n)$. If the most similar pair of classes to be amalgamated at some step of the algorithm is $\{C_r, C_s\}$, the height in the valued tree of their

Table 4.2 *Parameter values defining several clustering criteria incorporated in the Lance-Williams-Jambu recurrence relation.*

| | Clustering criterion | $\alpha_i$ | $\beta$ | $\gamma$ | $\delta_i$ | $\epsilon$ |
|---|---|---|---|---|---|---|
| C1 | Single link | $\frac{1}{2}$ | 0 | $-\frac{1}{2}$ | 0 | 0 |
| C2 | Complete link | $\frac{1}{2}$ | 0 | $\frac{1}{2}$ | 0 | 0 |
| C3 | Group average link | $\frac{n_i}{n_i+n_j}$ | 0 | 0 | 0 | 0 |
| C4 | Weighted average link | $\frac{1}{2}$ | 0 | 0 | 0 | 0 |
| C5 | Mean dissimilarity | $\dfrac{\binom{n_i+n_k}{2}}{\binom{n_+}{2}}$ | $\dfrac{\binom{n_i+n_j}{2}}{\binom{n_+}{2}}$ | 0 | $-\dfrac{\binom{n_i}{2}}{\binom{n_+}{2}}$ | $-\dfrac{\binom{n_k}{2}}{\binom{n_+}{2}}$ |
| C6 | Sum of squares | $\frac{n_i+n_k}{n_+}$ | $\frac{n_i+n_j}{n_+}$ | 0 | $\frac{-n_i}{n_+}$ | $\frac{-n_k}{n_+}$ |
| C7 | Incremental sum of squares | $\frac{n_i+n_k}{n_+}$ | $\frac{-n_k}{n_+}$ | 0 | 0 | 0 |
| C8 | Centroid | $\frac{n_i}{n_i+n_j}$ | $\frac{-n_in_j}{(n_i+n_j)^2}$ | 0 | 0 | 0 |
| C9 | Median | $\frac{1}{2}$ | $-\frac{1}{4}$ | 0 | 0 | 0 |

Note: $n_i$ denotes the number of objects in class $C_i$; $n_+ \equiv n_i + n_j + n_k$.

union, $h(C_r \cup C_s)$, is given by $d(C_r, C_s)$. If both $C_r$ and $C_s$ are singleton classes, this quantity is $d_{rs}$; otherwise, it is provided by equation (4.7).

Table 4.2 gives the values of the parameters $\theta$ that define several commonly used clustering criteria; additional criteria are listed by Podani (1989). The methodology is illustrated by analysing the data summarized in Tables 3.1 and 3.2 using the single link and incremental sum of squares criteria. Properties of the classifications provided by the clustering strategies defined in Table 4.2 are discussed later in the section.

Repeating the definition of single link classes given in Section 3.4.1: two objects, $i$ and $j$, belong to the same single link class at level $h$ if there exists a chain of $(m - 1)$ intermediate objects, $i_1, i_2, ..., i_{m-1}$, linking them such that

$$d_{i_k, i_{k+1}} \leq h \text{ for } k = 0, 1, ..., m - 1 \ (1 \leq m \leq n - 1), \qquad (4.8)$$

where $i_0 \equiv i$ and $i_m \equiv j$. Successively coarser partitions of the set of objects result as the value of $h$ is increased. In terms of an agglomerative algorithm (Sneath, 1957), two classes are merged when there is a single link between a pair of objects, one from each class, and

$$d(C_i \cup C_j, C_k) = \min(d(C_i, C_k), d(C_j, C_k)). \qquad (4.9)$$

As indicated in Table 4.2, this updating formula is the special case of the Lance-Williams-Jambu recurrence relation for which $\alpha_i = \frac{1}{2} = \alpha_j, \gamma = -\frac{1}{2}$, and all other parameter values are zero. The successive amalgamations for the dissimilarity matrix given in Table 3.2 are:

class $\{4, 5\}$ forms at height 98;

class $\{3, 6\}$ forms at height 149;

these two classes join, to form class $\{3 - 6\}$ at $(d_{46} =)$ 181;

object 7 joins class $\{3 - 6\}$, to form class $\{3 - 7\}$ at $(d_{67} =)$ 232;

object 2 joins class $\{3 - 7\}$, to form class $\{2 - 7\}$ at $(d_{23} =)$ 292;

object 1 joins class $\{2 - 7\}$, to form class $\{1 - 7\}$ at $(d_{14} =)$ 305.

A dendrogram summarizing this hierarchical classification is plotted in Fig. 4.6, the ordering of the leaves being chosen so that the dissimilarity matrix having rows and columns in this order is close to satisfying the Robinson property described in Section 4.1. The corresponding matrix of ultrametric distances is given in Table 4.3;

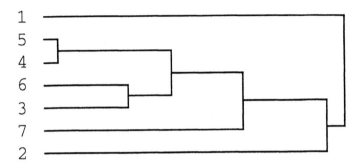

Figure 4.6 *A single link dendrogram of the set of seven objects whose dissimilarity matrix is given in Table 3.2.*

Table 4.3 *Ultrametric distances corresponding to the hierarchical classification shown in Fig. 4.5.*

| Object | | | | | | |
|---|---|---|---|---|---|---|
| 5 | 305 | | | | | |
| 4 | 305 | 98 | | | | |
| 6 | 305 | 181 | 181 | | | |
| 3 | 305 | 181 | 181 | 149 | | |
| 7 | 305 | 232 | 232 | 232 | 232 | |
| 2 | 305 | 292 | 292 | 292 | 292 | 292 |
| Object | 1 | 5 | 4 | 6 | 3 | 7 |

it can be seen that the elements of this matrix satisfy the ultrametric condition (4.1). Sectioning the single link dendrogram at any level $h$ produces the maximum split partition into single link classes at that level, as described in Section 3.4.1.

A single link analysis of the acoustic confusion data is summarized in Fig. 4.7. The results are similar to those depicted in Fig. 4.5, but fail to identify $\{za, zha\}$ as a separate class.

In the incremental sum of squares algorithm for the analysis of objects described by quantitative variables, an optimal amalgamation is defined to be one which leads to the minimum possible increase at that stage in the total sum of squared distances about the class centroids; there can of course be more than one pair of

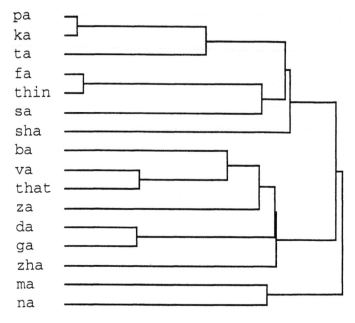

Figure 4.7 *A single link dendrogram of the acoustic confusion data.*

classes whose merging would lead to the same smallest increase in
the sum of squares. An early incremental sum of squares algorithm
was described by Ward (1963), and this method of analysis is often referred to as Ward's method. Anderson (1966) and Wishart
(1969b) showed that the problem could be expressed in terms of
the recurrence relation (4.7). Let $I(C_i, C_j)$ denote the increase in
the total sum of squared distances about the class centroids that
would result if classes $C_i$ and $C_j$, containing $n_i$ and $n_j$ objects
respectively, were amalgamated. If these two classes are merged,
then $I(C_i \cup C_j, C_k)$, the increase in the sum of squares that would
result from the merging of class $C_i \cup C_j$ and some other class $C_k$
(containing $n_k$ objects), is given by:

$$I(C_i \cup C_j, C_k) = \frac{n_i + n_k}{n_i + n_j + n_k} I(C_i, C_k)$$
$$+ \frac{n_j + n_k}{n_i + n_j + n_k} I(C_j, C_k)$$

$$- \quad \frac{n_k}{n_i + n_j + n_k} I(C_i, C_j). \qquad (4.10)$$

When $C_i = \{i\}$ and $C_j = \{j\}$, $I(C_i, C_j) = \frac{1}{2}d_{ij}$, where $d_{ij}$ denotes the squared Euclidean distance between the $i$th and $j$th objects. By multiplying through equation (4.10) by 2, one can thus define a measure of the dissimilarity between two *classes* of objects, and obtain a recurrence relation for evaluating such dissimilarities; as indicated in Table 4.2, this is the special case of the Lance-Williams-Jambu recurrence relation for which the non-zero parameter values are

$$\alpha_i = \frac{n_i + n_k}{n_i + n_j + n_k}, \ \alpha_j = \frac{n_j + n_k}{n_i + n_j + n_k}, \ \beta = \frac{-n_k}{n_i + n_j + n_k}.$$

This algorithm is illustrated by application to the set of squared distances given in Table 3.2. The successive modifications to the dissimilarity matrix are shown in the separate parts of Table 4.4. Thus, the first amalgamation is between objects 4 and 5, leading to an increase of $\frac{1}{2}(98)$ in the sum of squares. The dissimilarities between this new class $\{4, 5\}$ and each of the other objects are calculated using equation (4.7), to obtain the first matrix in Table 4.4. The smallest element of this dissimilarity matrix is $d_{36}$, shown starred, and so the second amalgamation leads to the formation of class $\{3, 6\}$ at height 149. Parts II - V of Table 4.4 show the modified dissimilarity matrices and starred smallest elements which define the third to sixth amalgamations. The set of amalgamations is summarized in the dendrogram shown in Fig. 4.8. The height of each class in the dendrogram is twice the increase in the sum of squares that results from the formation of that class by the merging of its two offspring classes; the total of the heights of all the internal nodes (98 plus the sum of the starred entries in Table 4.4) is 3596, twice the total sum of squares of the seven objects about their centroid (see Table 3.3).

Fig. 4.9 shows the hierarchical classification obtained when the incremental sum of squares algorithm is used to analyse the Abernethy Forest 1974 data. These data have already been used to illustrate stopping rules in Section 3.5, where it was concluded that there were five, or possibly seven, distinct classes in the data set. The five classes obtained by sectioning the dendrogram at the level shown in Fig. 4.9 correspond exactly to the five classes that were obtained when these data were analysed in Section 3.2.3.

Table 4.4 *Successive updates when the dissimilarity matrix given in Table 3.2 is analysed using the incremental sum of squares clustering criterion; the starred entries indicate the amalgamations to be carried out.*

| Part | Class | | | | | |
|------|-------|------|--------|--------|--------|-----|
| I | 2 | 757 | | | | |
| | 3 | 325 | 292 | | | |
| | (4, 5) | 654 | 1923.3 | 587.3 | | |
| | 6 | 634 | 785 | 149* | 283.3 | |
| | 7 | 1250 | 565 | 305 | 1184.7 | 232 |
| | Class | 1 | 2 | 3 | (4, 5) | 6 |
| II | 2 | 757 | | | | |
| | (3, 6) | 589.7 | 668.3 | | | |
| | (4, 5) | 654 | 1923.3 | 578.5 | | |
| | 7 | 1250 | 565 | 308.3* | 1184.7 | |
| | Class | 1 | 2 | (3, 6) | (4, 5) | |
| III | 2 | 757 | | | | |
| | (3, 6, 7) | 990.2 | 706.7 | | | |
| | (4, 5) | 654* | 1923.3 | 1050.3 | | |
| | Class | 1 | 2 | (3, 6, 7) | | |
| IV | 2 | 1657.5 | | | | |
| | (3, 6, 7) | 1208.3 | 706.7* | | | |
| | Class | (1, 4, 5) | 2 | | | |
| V | (2, 3, 6, 7) | 1680* | | | | |
| | Class | (1, 4, 5) | | | | |

Some comments follow on properties of the clustering criteria defined in Table 4.2. In discussing procedures $C1 - C5$, it is convenient to consider a graph theoretic representation in which each object is represented by a vertex in the graph and an edge links vertices $i$ and $j$ at level $h$ if the $i$th and $j$th objects have dissimilarity $d_{ij} \leq h$. For a specified value of $h$, single link classes are connected components of the graph; by varying the value of $h$, the complete set of nested single link classes is obtained. From their definition, single link classes are isolated from one another but need not possess much internal cohesion. They also have the

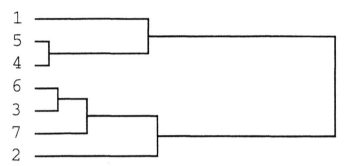

Figure 4.8 *Incremental sum of squares dendrogram for the set of seven objects described in Table 3.1.*

property that small changes in the dissimilarity matrix do not lead to large changes in the ultrametric matrix defining the hierarchical classification (Jardine and Sibson, 1971, Chapter 9).

The complete link algorithm (McQuitty, 1960) defines the dissimilarity between two classes $C_i$ and $C_j$ by

$$d(C_i, C_j) \equiv \max_{r,s}(d_{rs} \mid r \in C_i, s \in C_j) \qquad (4.11)$$

Complete link classes are (some, but not necessarily all, of the) complete subgraphs of the set of graphs resulting from varying the value of $h$. The partition into $c$ classes obtained by sectioning a complete link dendrogram may be regarded as an approximation to the partition into $c$ classes that minimizes the maximum diameter criterion $P(H3, Max)$ defined in Section 3.1. An algorithm for obtaining a partition into a specified number of classes that minimizes criterion $P(H3, Max)$ is presented by Hansen and Delattre (1978); the optimal partitions into different numbers of classes need not be hierarchically nested, and the complete link algorithm cannot be guaranteed to find them. From their definition, complete link classes are compact but need not be externally isolated.

The single link and complete link clustering criteria take complementary approaches to the aim that classes be internally cohesive and externally isolated, each of them concentrating on satisfying one of these desiderata. The average link procedures were designed to take a middle road between these extremes, the dissimilarity between two classes being defined to be the average ($C3$: Sokal and Michener, 1958) or weighted average ($C4$: McQuitty, 1966, 1967)

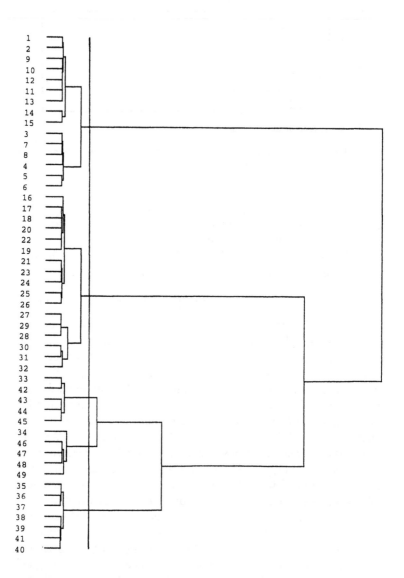

Figure 4.9 *Incremental sum of squares dendrogram of the Abernethy Forest 1974 data.*

of the pairwise dissimilarities between objects in different classes. Criterion $C5$ (Podani, 1989) defines the height of a class to be the average within-class dissimilarity; sectioning such a dendrogram provides a partition that may be regarded as an approximation to a partition which minimizes the largest average within-class dissimilarity.

Clustering procedures $C6-C9$ implicitly assume that the objects can be represented by points in some space and that the measure of dissimilarity between a pair of objects is the squared Euclidean distance between the corresponding pair of points. The dissimilarity between classes $C_i$ and $C_j$ is defined to be the sum of the squared distances about the centroid of $C_i \cup C_j$ by criterion $C6$ (Jambu, 1978), and the *increase* in the sum of squares that would result from the fusion of $C_i$ and $C_j$ by criterion $C7$ (Ward, 1963). One could regard the partitions into c classes provided by sectioning each of these dendrograms to be approximations to the partitions into $c$ classes that minimize, respectively, the largest within-class sum of squares $(P(H1, Max))$, and the total within-class sum of squares $(P(H1, \Sigma))$. Wright (1973) presents a list of axioms that are satisfied by this latter criterion.

The dissimilarity between classes $C_i$ and $C_j$ is defined to be the squared distance between their centroids by criterion $C8$ (Sokal and Michener, 1958; Gower, 1967); criterion $C9$ (Lance and Williams, 1966b; Gower, 1967) modifies this definition by assigning each class the same weight in calculating the 'centroid'. It seems more difficult to state any property possessed by a partition obtained by sectioning such dendrograms, but a partition obtained from application of the centroid strategy might be regarded as an approximation to a partition with the property that the sum of the squares of the distances between each pair of centroids is minimized.

The centroid and median procedures can lead to 'reversals' in the 'hierarchical classification', i.e. it can contain classes $C_i$ and $C_j$ for which

$$C_i \subset C_j \text{ but } h(C_i) > h(C_j).$$

Necessary and sufficient conditions for the absence of reversals in hierarchical classifications produced by Lance and Williams's (1966b, 1967) recurrence relation are (Milligan, 1979; Batagelj, 1981):

$$\gamma \geq -\min(\alpha_i, \alpha_j) \tag{4.12}$$

$$\alpha_i + \alpha_j \geq 0 \tag{4.13}$$

$$\alpha_i + \alpha_j + \beta \geq 1. \tag{4.14}$$

Lance and Williams (1967) introduced the concept of space distortion, referring to clustering procedures as space-conserving, space-contracting and space-dilating. These ideas were formalized by Dubien and Warde (1979). Assuming that

$$d(C_i, C_j) < d(C_i, C_k) < d(C_j, C_k),$$

a clustering strategy is defined to be space-conserving if

$$d(C_i, C_k) < d(C_i \cup C_j, C_k) < d(C_j, C_k), \tag{4.15}$$

and to be space-contracting if the first inequality is broken, and space-dilating if the second inequality is broken. Single link is space-contracting and complete link is space-dilating. Strategies that can lead to reversals can be defined to be 'highly space-contracting', since one can have

$$d(C_i \cup C_j, C_k) < d(C_i, C_j).$$

Discussions of the space-conserving and space-distorting properties of various combinations of parameter values are provided by Dubien and Warde (1979) and Ohsumi and Nakamura (1989). Chen and Van Ness (1996) present necessary and sufficient conditions under which Lance and Williams's (1966b, 1967) recurrence relation is space-conserving, space-dilating and space-contracting.

The Lance-Williams-Jambu recurrence relation includes many commonly used clustering criteria and has the advantage that information about individual objects need not be retained after they have been incorporated into larger classes. However, there are many algorithms that do not fit into this framework, including ones based on the proportion of links present between classes (Kendrick and Proctor, 1964; Sneath, 1966; Day and Edelsbrunner, 1985) and other graph-theoretic concepts (Ling, 1972, 1973a; Hubert, 1974). There is also the matter of the efficiency of algorithms: thus, a simple program implementing the Lance-Williams-Jambu recurrence relation would have $O(n^3)$ time complexity and $O(n^2)$ space complexity. By associating with each class a priority queue that orders the other classes by their proximity to it, the time complexity can be reduced to $O(n^2 \log n)$ (Day and Edelsbrunner, 1984), and further improvements in efficiency are possible for some clustering criteria using the concept of *reducibility* (Bruynooghe, 1978).

Classes $C_i$ and $C_j$ are defined to be *reciprocal nearest neighbours* (McQuitty, 1960) if

$$d(C_i, C_j) \leq \min(d(C_i, C_k), d(C_j, C_k)) \text{ for all } k \neq i, j. \quad (4.16)$$

The reducibility condition holds if, for all $i, j, k$,

$$d(C_i, C_j) \leq \min(d(C_i, C_k), d(C_j, C_k))$$
$$\Rightarrow \min(d(C_i, C_k), d(C_j, C_k)) \leq d(C_i \cup C_j, C_k), \quad (4.17)$$

i.e. the amalgamation of classes $C_i$ and $C_j$ cannot lead to a class $C_i \cup C_j$ which is closer to a class $C_k$ than either $C_i$ or $C_j$ is, thus ruling out the possibility of obtaining any reversals in the classification. Of the clustering criteria listed in Table 4.2, only $C8$ and $C9$ cannot be guaranteed to satisfy the reducibility condition. By defining $\tau$ such that

$$d(C_i, C_j) \leq \tau \leq \min(d(C_i, C_k), d(C_j, C_k))$$

it follows from (4.17) that

$$\tau \leq d(C_i \cup C_j, C_k).$$

Hence, for clustering criteria that satisfy the reducibility condition, such as $C1 - C7$, all reciprocal nearest neighbours can be amalgamated at the same time without altering the later stages of the agglomeration. This property extends the size of data set that can be analysed, since one does not need to store the entire dissimilarity matrix but only dissimilarities less than a threshold value; by successively increasing the value of the threshold, the entire hierarchical classification can be efficiently constructed.

Efficient algorithms for identifying reciprocal nearest neighbours have made use of *nearest neighbour chains* (Benzécri, 1982; Juan, 1982; Murtagh, 1983): if $NN(i)$ denotes the nearest neighbour of node $i$, a nearest neighbour chain is a sequence $(i_0, (i_r = NN(i_{r-1})$ $(r = 1, ..., k)), i_{k-1} = NN(i_k))$ that ends in a pair of reciprocal nearest neighbours $(i_{k-1}, i_k)$. Each node represents a class, initially a singleton class. Starting from node $i_0$, reciprocal nearest neighbours $i_{k-1}$ and $i_k$ are identified and joined into a single node; thereafter, the chain continues from $i_{k-2}$ if $k \geq 2$, or a new chain is started.

Hierarchical classifications have also been obtained using parallel computers (Li and Fang, 1989; Rasmussen and Willett, 1989; Li, 1990; Olson, 1995; Tsai et al., 1997). Considerable savings in time can be achieved in the implementation of some algorithms.

### 4.2.3 Other algorithms

Other algorithms for obtaining hierarchical classifications include some that can be categorized as 'incremental' or 'divisive'. In incremental algorithms, a classification of $m$ objects is augmented by successively inserting new objects into the classification ($m = 2, ..., n-1$). The lower triangular dissimilarity matrix is read in one row at a time; thus, storage requirements are only $O(n)$. The updating can be conveniently carried out by making use of the representation defined for the right-justified dendrogram Fig 4.2(c), in which each object is described by two functions: the rightmost member of the class that it joins when it ceases to be the rightmost member of its own class, and the height at which this occurs. This approach has been used to provide algorithms with $O(n^2)$ time complexity for obtaining hierarchical classifications based on the single link (Sibson, 1973) and complete link (Defays, 1977) clustering criteria.

At the start of divisive algorithms, all objects belong to a single class. The number of classes is increased at each step of the algorithm by dividing an existing class into (usually, two) sub-classes. The problem of finding an optimal division has been shown to be NP-hard for several clustering criteria (Brucker, 1978; Welch, 1982), but polynomial-time algorithms exist for finding a partition that minimizes the larger of the two sub-class diameters (Rao, 1971; Hubert, 1973b; Guénoche, Hansen and Jaumard, 1991) and the sum of the two sub-class diameters (Hansen and Jaumard, 1987). As with agglomerative algorithms, these divisive algorithms are stepwise optimal, and one cannot guarantee that any optimality properties are possessed by the complete hierarchical classification or a partition into more than two classes. However, interest usually resides mostly in the larger classes in a classification, and it can be argued that when divisive algorithms are computationally feasible they are preferable to agglomerative algorithms which provide larger classes only after a number of stepwise optimal amalgamations. Further algorithms are discussed by Gordon (1996a).

### 4.2.4 Parsimonious trees

The hierarchical classifications produced by clustering algorithms usually comprise partitions into $c$ classes for all values of $c$ between 1 and $n$, being represented by dendrograms that contain $(n - 1)$ internal nodes. The complete sets of partitions and classes do not appear to be used by investigators, and can hinder interpretation. One approach to resolving this difficulty has involved the construction of *parsimonious trees* which contain a limited number of internal nodes; some information is discarded in this process, but the main features of the data might be represented more clearly. Such trees can be constructed to have one of the following properties:

(i) they contain a specified number, $c_1$, of internal nodes

(ii) the heights of the nodes take a specified number, $c_2$, of different values.

Algorithms have been proposed for seeking parsimonious trees directly from (dis)similarity data (Hartigan, 1967; Sriram, 1990) and by simplifying complete hierarchical classifications (Gordon, 1987); this latter work is described below. The aim is to obtain a partition of the internal nodes that 'respects' the original classification. Each class in this partition defines one or more *favoured* nodes; these are the only nodes that are retained in the parsimonious tree, each other node being combined with the first favoured node encountered on the path between it and the root. The methodology is illustrated by application to the hierarchical classification depicted in Fig. 4.1. A truncated version of this classification, restricted to illustrating the relationships between the internal nodes $A - G$, is shown in Fig. 4.10($a$).

Two different types of parsimonious tree are described, differing in whether or not it is relevant to make comparisons between non-nested subsets of objects, such as $B$ and $C$ in Fig. 4.10($a$). If such comparisons are not considered appropriate, attention is restricted to making 'local' comparisons between internal nodes corresponding to nested subsets. By removing $(c_1 - 1)$ edges from the tree linking the internal nodes, one obtains a partition of these inter-

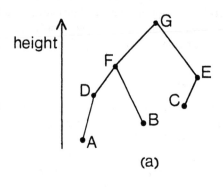

(a)

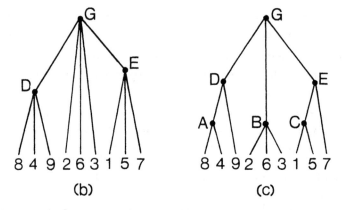

(b)                                              (c)

Figure 4.10 *Illustrating definitions of parsimonious trees: (a) a tree in-dicating the relationship between the seven internal nodes that define the classification of the set of nine objects depicted in Fig. 4.1; (b) a local parsimonious tree of the nine objects; (c) a global parsimonious tree of the nine objects.*

nal nodes into $c_1$ classes. In each of these classes, the node that is closest to the root is designated a favoured node; by definition, the root itself is always a favoured node. The tree comprising only the favoured nodes and the singleton subsets, containing only $c_1$ inter-nal nodes, is called a *local* parsimonious tree. For example, if one removes edges $DF$ and $EG$ from Fig. 4.10(a), the internal nodes are partitioned into the three classes $\{A, D\}, \{B, F, G\}$ and $\{C, E\}$, the favoured nodes are $D, G$ and $E$, and the local parsimonious tree is as shown in Fig. 4.10($b$).

If it is relevant to compare the heights of non-nested subsets of objects, one can ensure that there are only $c_2$ different values for the heights of the internal nodes by drawing $(c_2 - 1)$ horizontal lines in some of the $(m - 1)$ gaps between the $m$ internal nodes, and requiring all internal nodes lying between two neighbouring lines (or above the highest line or below the lowest line) to have the same height. For example, if two horizontal lines are drawn in Fig. 4.10($a$), just above internal nodes $C$ and $E$, the nodes are partitioned into the three classes $\{A, B, C\}, \{D, E\}$ and $\{F, G\}$, all nodes in the same class having the same height. The corresponding *global* parsimonious tree is depicted in Fig. 4.10($c$).

A hierarchical classification, with ultrametric distances $(h_{ij})$ can be transformed into a parsimonious tree, with ultrametric distances $(p_{ij})$, so as to minimize a measure of distortion, such as a weighted sum of squared distances,

$$\Sigma_{1 \leq j < i \leq n} w_{ij} (h_{ij} - p_{ij})^2,$$

subject to conditions which ensure that $(p_{ij})$ defines either a local parsimonious tree with the required number of internal nodes, or a global parsimonious tree with the required number of different node heights; fuller details are provided by Gordon (1987). The algorithms for carrying out the transformation are examples of constrained classification algorithms, which are discussed more fully in Section 5.2.

Global parsimonious trees appear particularly relevant for taxonomic investigations in which one wishes to identify different ranks in a hierarchy; local parsimonious trees appear more relevant for the exploratory analysis of data, when it is quite possible for different classes to possess different degrees of internal homogeneity and hence different heights in the valued tree.

As an illustration of this methodology, Fig. 4.11 shows a group average link dendrogram of the European fern data, and Fig. 4.12 shows a local parsimonious tree containing only ten internal nodes. It is interesting to note the extent to which geographically contiguous regions cluster together: some of the internal nodes correspond to regions from north Russia (709 - 713), the rest of north-east Europe, the southern Mediterranean (101 - 106), and the rest of

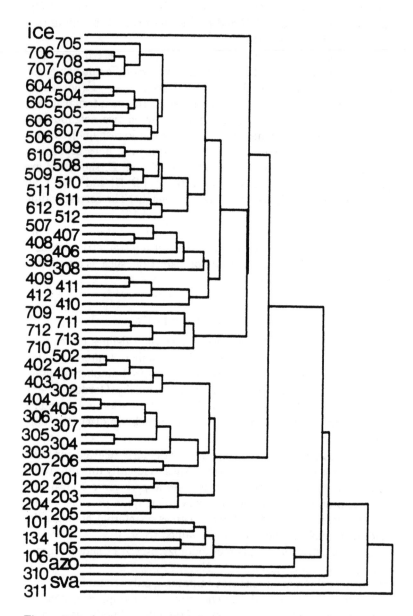

Figure 4.11 *A group average link dendrogram of the European fern data. Reproduced from Gordon (1987).*

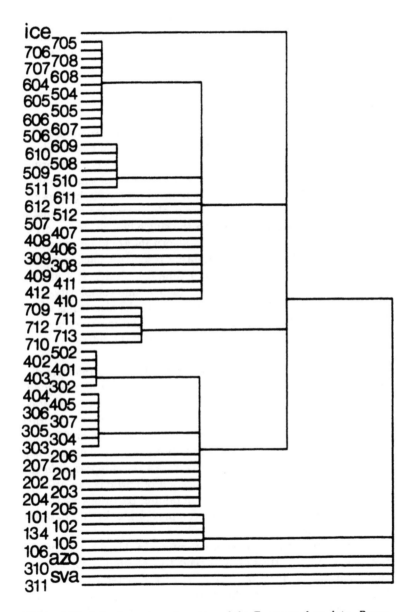

Figure 4.12 *A local parsimonious tree of the European fern data. Reproduced from Gordon (1987).*

south-west Europe; two regions from south-east Europe (310, 311) and the three islands are seen to be distinct from other regions. Although these features are present in the original classification displayed in Fig. 4.11, they appear much more clearly in the parsimonious tree.

## 4.3 Choice of clustering strategy

In the previous section, it was convenient to categorise material by the type of algorithm used to obtain a hierarchical classification. However, it is the properties of the classification that are of prime importance, and the only relevant criterion for assessing an algorithm is the efficiency with which it provides the desired result. For example, there are many different algorithms for obtaining a hierarchical classification based on the single link clustering criterion (Rohlf, 1982a) and a preferred algorithm would be one which makes minimal demands on computing resources; thus, given the existence of algorithms with $O(n^2)$ time complexity, it would be inefficient to use an algorithm with higher time complexity. This distinction between clustering 'methods' and the algorithms which implement them has been stressed by many authors, notably Jardine and Sibson (1971, Chapter 6), but is still the subject of some confusion.

A property of numerical classifications that is particularly disconcerting to newcomers to the field is the fact that markedly different results can be obtained when a set of objects is analysed using different clustering procedures. The point is that clustering criteria are not model-free: as illustrated in Figure 3.5 and Section 3.6, clustering criteria implicitly specify models for data and can provide misleading summaries of the class structure present in the data. This section addresses the question of how one can attempt to select methods of analysis (for obtaining partitions as well as hierarchical classifications) that are appropriate for the data set under consideration.

When defining various clustering criteria in this chapter and the previous chapter, attention was paid to describing properties of the resulting classification. For most of the criteria presented in Section 4.2.2, it was possible only to describe properties of the constituent partitions. For a complete hierarchical classification, it is more natural to seek a transformation from a set of pairwise dissimilarities $(d_{ij})$ to a set of ultrametric distances $(h_{ij})$ that min-

Table 4.5 *Selected measures of discordance and agreement between dissimilarities $(d_{ij})$ and ultrametric distances $(h_{ij})$.*

| $D1$ | Weighted sum of squares | $\Sigma w_{ij}(d_{ij} - h_{ij})^2$ |
|---|---|---|
| $D2$ | Minkowski metrics | $\begin{cases} \{\Sigma\mid d_{ij} - h_{ij}\mid^{1/\lambda}\}^{\lambda} \ (0 < \lambda \leq 1) \\ \max_{\{i,j\in\Omega\}}\mid d_{ij} - h_{ij}\mid \ (\lambda = 0) \end{cases}$ |
| $\overline{D3}$ | Correlation coefficient | $\dfrac{\Sigma(d_{ij}-\bar{d})(h_{ij}-\bar{h})}{[\Sigma(d_{ij}-\bar{d})^2\Sigma(h_{ij}-\bar{h})^2]^{\frac{1}{2}}}$ |
| $\overline{D4}$ | Goodman-Kruskal $\gamma$ | $(S_+ - S_-)/(S_+ + S_-)$, where $S_+$ (resp., $S_-$) denotes the number of concordant (resp., discordant) pairs $\{(i,j),(k,l)\}$. |

imizes some measure of the discordance between $(d_{ij})$ and $(h_{ij})$. Table 4.5 presents two measures of discordance, $D1$ and $D2$, and two measures of agreement, $\overline{D3}$ and $\overline{D4}$. The least squares measure $D1$ was discussed in Section 4.2.1. If attention is restricted to obtaining ultrametric distances that satisfy

$$h_{ij} \leq d_{ij} \text{ for all } i, j \in \Omega,$$

and the measure of discordance is $D2$, Jardine and Sibson (1971) prove that the single link method is the unique clustering method satisfying a stated list of axioms. This is a valuable characterization of the single link method that allows investigators to examine Jardine and Sibson's (1971) axioms to establish if they would be appropriate for the analysis of their data. Few investigators would analyse data using only the single link method of analysis. The major criticism of the axioms has centred on the continuity condition that small changes in $(d_{ij})$ should be guaranteed for all data sets to lead to only small changes in $(h_{ij})$; a preferable solution is to check if large changes occur for the data set under investigation (Cormack, 1971; Gower, 1971b).

It has also been observed empirically that (restricting attention to a subset of common clustering criteria) the criterion which maximizes the (cophenetic) correlation coefficient ($\overline{D3}$: Sokal and

Rohlf, 1962) is often group average link, $C3$ (Sokal and Rohlf, 1962; Sneath, 1969). However, such results merely reformulate the problem of specifying a relevant clustering criterion in terms of defining an appropriate measure of the distortion of the transformation from the data to the required classification. Further, it seems to be of dubious validity to compare dissimilarities and ultrametric distances by measuring the strength of their *linear* relationship, particularly given the fact that a set of ultrametric distances contains many tied values.

In measure $\overline{D4}$, comparisons are made between all pairs of pairs of objects. The comparison between pairs $(i, j)$ and $(k, l)$ can be described by

$$d_{ij} \, R_d \, d_{kl} \text{ and } h_{ij} \, R_h \, h_{kl},$$

where the relation symbols $R_d, R_h \in \{<, >\}$. A comparison is said to be concordant if $(R_d, R_h) = (<, <)$ or $(>, >)$, and to be discordant if $(R_d, R_h) = (<, >)$ or $(>, <)$; $\overline{D4}$ is based on the total numbers of concordant and discordant pairs, as defined in Table 4.5. This measure has the advantage that no comparisons are made when there are tied values.

To assist in their choice of clustering criteria, investigators draw as much information as they can from the background to the problem about the nature of the data and the manner in which it would be useful to represent them. Help may be provided by considering the relevance (for the data being investigated) of the axioms given by Jardine and Sibson (1971) and Wright (1973). A knowledge of properties of various clustering procedures, such as the relative importance they attach to the concepts of class isolation and cohesion, might allow specification of appropriate methods of analysis. This approach is more fully developed in the admissibility work of Fisher and Van Ness (1971). These authors state various properties which one might expect 'reasonable' clustering procedures, or the results obtained from them, to possess. If $A$ denotes a desired property, then any clustering procedure which satisfies $A$ is called $A$-admissible. Some of the properties introduced by Fisher and Van Ness (1971) are given below:

*Convex admissibility:* If the objects can be represented by a set of points in some Euclidean space, a partition into $c$ classes $\{C_i \, (i = 1, ..., c)\}$ is said to be convex admissible if the convex hulls of

Table 4.6 *Admissibility table, indicating whether or not selected clustering procedures satisfy several admissibility conditions.*

| Clustering procedure | Admissibility condition | | |
|---|---|---|---|
| | Convex admissibility | Point proportion admissibility | Monotone admissibility |
| Single link | No | Yes | Yes |
| Complete link | No | Yes | Yes |
| Sum of squares | Yes | No | No |
| Group average link | No | No | No |

$C_1, C_2, ..., C_c$ do not intersect. A convex admissible clustering procedure is one which can be guaranteed to provide convex admissible partitions for all data sets.

*Point proportion admissibility*: A clustering procedure is said to be point proportion admissible if the same class boundaries are obtained when one uses it to analyse any data set and a modified version of that data set in which one or more objects have been replicated any number of times.

*Monotone admissibility*: A clustering procedure is said to be monotone admissible if applying a monotone transformation to the elements of the (dis)similarity matrix does not alter the classes in a partition or the ranked tree of a hierarchical classification.

Table 4.6 summarizes whether or not several clustering procedures satisfy these three admissibility conditions; note that these procedures include criteria defining an optimal solution (single link, sum of squares) and stepwise optimal agglomerative clustering algorithms (complete link, group average link). More extensive tables, containing further clustering procedures and admissibility conditions, are given by Fisher and Van Ness (1971), Van Ness (1973) and Chen and Van Ness (1994).

Such Tables are used as follows. Suppose that:

(i) the data were collected by some sampling mechanism and

there was the possibility that some objects had been recorded more than once;

(ii) there was some uncertainty that the measure of dissimilarity was an accurate reflection of differences between the objects, but confidence was higher that the dissimilarities were ranked in the correct order.

In these circumstances, one might wish to use a clustering procedure that is both point proportion admissible and monotone admissible. Table 4.6 indicates that the sum of squares and group average link clustering procedures are inadmissible, i.e. inappropriate *for these particular data*, no claims being made concerning their relevance for analysing other data sets. If restricting attention to the clustering procedures included in Table 4.6, one would analyse the data using either single link or complete link, or preferably both.

In general, if it is not possible to determine a single preferred clustering procedure, it is useful to analyse data using two or more 'sensible' methods of analysis and synthesize the results. This can involve superimposing classifications onto graphical representations of the data and/or obtaining consensus classifications. If very similar results are obtained from separate analyses, one can be more confident that inappropriate structure is not being imposed on the data. The next section describes ways of defining and obtaining consensus trees.

It is also valuable to validate the results of a classification study, a topic that is described in Section 7.2.

## 4.4 Consensus trees

Given $t$ ($\geq 2$) hierarchical classifications $\{T_r \ (r = 1, ..., t)\}$ of the same set of $n$ objects, it is relevant to obtain a single *consensus tree* which synthesizes the information contained in the original classifications and might therefore provide a more reliable summary of the relationships between the objects. The original classifications might have been obtained by analysing the objects using $t$ different clustering procedures, or might have been specified directly by $t$ different investigators or by a single investigator considering $t$ different sets of variables describing the objects; in this latter case, there has been discussion about whether or not it would be more appropriate to analyse a single data set comprising objects described by the combined set of variables (e.g. Barrett, Donoghue and Sober, 1991; de Queiroz, 1993)

Most of consensus trees methodology is concerned with the synthesis of $n$-trees and thus considers only the constituent classes of each tree. This work can be categorized by whether it is relevant to consider only exact correspondence of classes in different trees or whether partial agreement of class memberships should be taken into account. The first category includes strict consensus trees and majority rule consensus trees, and the second category includes various types of 'intersection' tree.

A *strict consensus tree* (Sokal and Rohlf, 1981) comprises those classes that belong to all of the $t$ original trees. Strict consensus trees can contain few classes, and proposals have been made to augment them by the addition of some extra compatible classes (e.g. Nelson, 1979; Bremer, 1990). McMorris, Meronk and Neumann (1983) require each class in a consensus tree to belong to at least $m$ ($[t/2] + 1 \leq m \leq t$) of the original trees. The case $m = [t/2] + 1$ provides the *majority rule consensus tree* (Margush and McMorris, 1981). Barthélemy (1988) considers consensus $n$-trees defined by two numbers $(m_1, m_2)$, such that a class belonging to $k$ of the original $n$-trees belongs to every consensus $n$-tree if $k \geq m_2$, no consensus $n$-tree if $k < m_1$, and might belong to a consensus $n$-tree if $m_1 \leq k < m_2$.

These concepts are illustrated in Fig. 4.13, the trees in which should all be regarded as $n$-trees, i.e. the heights of the internal nodes are irrelevant. The strict consensus tree of $T_1$ and $T_2$ is $T_4$, but the strict consensus tree of $T_r$ and $T_3$ (for $r = 1$ or 2) is the bush comprising only the singleton classes and the class containing all the objects. The majority rule consensus tree of $T_1, T_2$ and $T_3$ is $T_5$, since each of its constituent classes belongs to two of $\{T_1, T_2, T_3\}$.

A strict consensus $n$-tree of the hierarchical classifications of the acoustic confusion data depicted in Figs. 4.5 and 4.7 contains the following classes, in addition to the class of all objects and the singleton classes: $\{pa, ka\}$, $\{pa, ka, ta\}$, $\{fa, thin\}$, $\{fa, thin, sa\}$, $\{pa, ka, ta, fa, thin, sa\}$, $\{pa, ka, ta, fa, thin, sa, sha\}$, $\{va, that\}$, $\{ba, va, that\}$, $\{da, ga\}$, $\{ba, va, that, da, ga, za, zha\}$, $\{ma, na\}$. These results are shown in Fig. 4.14, but no significance should be attached to the heights of the classes in this Figure, since only subset information was used in obtaining the consensus tree.

A *median consensus tree* is an $n$-tree $T_M$ which minimizes

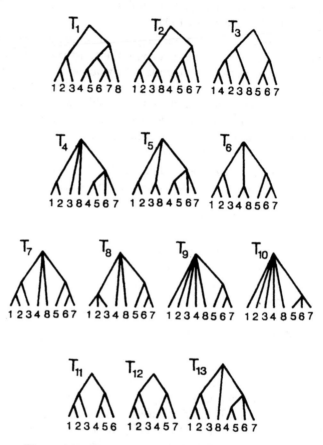

Figure 4.13 *Illustrating definitions of consensus trees.*

$$\Sigma_{r=1}^{t} D_s(T_r, T_M),$$

where $D_s(T_r, T_M)$ is defined to be the number of classes that belong to precisely one of $\{T_r, T_M\}$. If $t$ is odd, the majority rule consensus tree is the (unique) median consensus tree (Margush and McMorris, 1981; Bandelt and Barthélemy, 1984). If $t$ is even, there can be more than one median consensus tree, but each of them contains the classes belonging to the majority rule consensus tree together with some classes that belong to precisely half of the original trees (Barthélemy and McMorris, 1986).

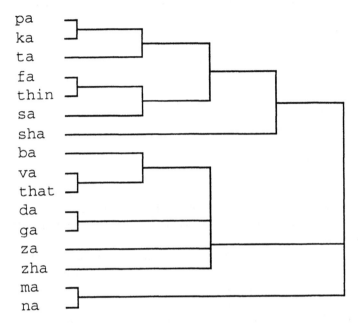

Figure 4.14 *The strict consensus tree of the trees depicted in Figs. 4.5 and 4.7.*

The second category of consensus trees involves obtaining the $t$-fold intersection of classes obtained by sectioning each tree at the same value of a 'generalized height', where definitions of the generalized height of a class include the number of objects it contains (Neumann, 1983). Stinebrickner (1984) defined the *consensus strength*, $s$, of a class $C$ to be

$$\max\{\text{card}(C)/\text{card}(\cup_{r=1}^{t} C_r) \mid C \subseteq C_r \in T_r \ (r = 1, ..., t)\}, \quad (4.18)$$

and the *s-consensus tree* to comprise solely those classes whose strength is at least $s$. This defines a family of consensus trees that contain a non-increasing set of classes as $s$ increases from 0 to 1; the 0-consensus tree is the intersection consensus tree of Neumann (1983) and the 1-consensus tree is the strict consensus tree if the generalized height of a class is defined to be the number of objects it contains. For example, when trees $T_1$ and $T_3$ are compared, the strengths of the classes defined at various values of the height, $h$, are:

$h = 3 : s(\{1,2\}) = \frac{2}{4}, s(\{6,7\}) = \frac{2}{3}$;

$h = 4 : s(\{5,6,7\}) = \frac{3}{4}$;

$h = 5 : s(\{4,8\}) = \frac{2}{8}, s(\{1,2,3\}) = \frac{3}{5}$.

The corresponding family of $s$-consensus trees is:

$T_6$ for $0 \leq s \leq 1/4$;

$T_7$ for $1/4 < s \leq 1/2$;

$T_8$ for $1/2 < s \leq 3/5$;

$T_9$ for $3/5 < s \leq 2/3$;

$T_{10}$ for $2/3 < s \leq 3/4$;

a bush  for $3/4 < s \leq 1$.

Much attention has been paid to obtaining axiomatic charac-terizations of consensus functions and investigating if they possess certain desirable properties; a recent review of this material is pro-vided by Leclerc (1998).

The consensus trees that are described above all contain the same leaves as each of the original trees. An alternative approach to consensus involves pruning as few leaves and attached branches as possible from each of the trees so as to obtain equivalent $n$-trees (Rosen, 1978; Gordon, 1980) or ranked trees (Gordon, 1979). The resulting *largest common pruned trees* or *maximum agreement subtrees* need not be uniquely defined: for example, $T_{11}$ and $T_{12}$ are both largest common pruned $n$-trees of $T_1$ and $T_2$, obtained by removing from them object 8 and either object 7 or object 6. An alternative approach provides a set of *reduced* trees that can contain objects in common (Wilkinson, 1994, 1995). Efficient algorithms for finding largest common pruned trees of two $n$-trees are presented by Steel and Warnow (1993) and Goddard et al. (1994); Amir and Keselman (1997) describe an algorithm for finding a largest common pruned tree of $t(> 2)$ $n$-trees when the degree of at least one of the trees is bounded.

Pruned branches can also be regrafted back on to common pruned trees, by inserting them at the lowest level for which their relation-ships with objects in the pruned trees do not contradict any of the original $t$ trees (Gordon, 1980); algorithmic considerations are dis-cussed by Goddard et al. (1995). If regrafting is carried out on a largest common pruned $n$-tree, the pruned and regrafted $n$-tree (which need not be uniquely defined) contains all of the classes in the strict consensus tree and possibly some other classes (Finden and Gordon, 1985); for example, the pruned and regrafted consensus $n$-tree of $T_1$ and $T_2$ is $T_{13}$, which contains the class $\{1, 2, 3\}$ in addition to all of the classes in the strict consensus tree $T_4$.

Consensus valued trees have been obtained by synthesizing the information contained in $t$ matrices of ultrametric distances, $\{(h_{ijr})$ $(i, j = 1, ..., n; r = 1, ..., t)\}$ (Ghashgai, Stinebrickner and Suters, 1989; Lapointe and Cucumel, 1991). The problem can be formulated in terms of finding ultrametric distances $\mathbf{u} \equiv (u_{ij})$ which minimize

$$L_2(\mathbf{u}) \equiv \Sigma_{r=1}^{t}\Sigma_{1\leq j<i\leq n}(h_{ijr} - u_{ij})^2. \qquad (4.19)$$

Since

$$L_2(\mathbf{u}) = \Sigma_{r=1}^{t}\Sigma_{1\leq j<i\leq n}(h_{ijr} - \bar{h}_{ij})^2 + t\Sigma_{1\leq j<i\leq n}(\bar{h}_{ij} - u_{ij})^2, \qquad (4.20)$$

where

$$\bar{h}_{ij} \equiv \Sigma_{r=1}^{t}h_{ijr}/t \ (1 \leq j < i \leq n), \qquad (4.21)$$

minimizing $L_2(\mathbf{u})$ is equivalent to minimizing

$$\Sigma_{1\leq j<i\leq n}(\bar{h}_{ij} - u_{ij})^2. \qquad (4.22)$$

This can be achieved using the penalty function approach described in Section 4.2.1. For example, Fig. 4.15 shows a dendrogram of the least squares consensus valued tree of the two classifications of the acoustic confusion data presented in Figs. 4.5 and 4.7.

Thus far, it has been assumed that each of the $t$ original trees specifies a hierarchical classification of the same set of $n$ objects. Trees that have some but not all of their leaves in common can be synthesized, to provide a consensus *supertree*, defining a classification of the complete set of objects. For $n$-trees, relevant methodology is presented by Gordon (1986b). Even if the original $n$-trees do not contradict one another, there can be many different supertrees; Steel (1992) presents necessary and sufficient conditions for there to be a unique solution, and Constantinescu and Sankoff (1995) describe an efficient algorithm for obtaining all binary supertrees.

Valued trees based on overlapping but not identical sets of objects can also be synthesized. Brossier (1990) presents an algorithm for merging separate ultrametric matrices and describes necessary and sufficient conditions for the solution to be uniquely defined. Lapointe and Cucumel (1995) obtain a least squares consensus valued tree by obtaining the ultrametric distances $(u_{ij})$ that minimize

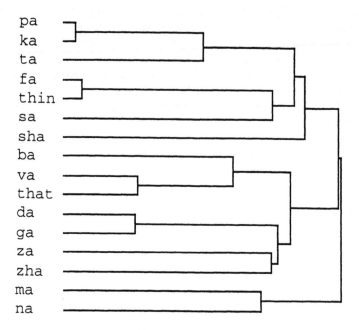

Figure 4.15 *The least squares consensus dendrogram of the dendrograms depicted in Figs. 4.5 and 4.7.*

Expression (4.22), where now the elements in the matrix $(\bar{h}_{ij})$ are averaged over different numbers of valued trees and missing values require to be estimated.

Throughout this section, the emphasis has been on defining and obtaining consensus trees. Tests of the validity of such trees can also be carried out, using methodology described in Section 7.2.5.

## 4.5 More general tree models

In obtaining a valued tree from a matrix of the pairwise dissimilarities within a set of objects, it is implicitly assumed that the dissimilarities $(d_{ij})$ and the ultrametric distances $(h_{ij})$ defining the valued tree can be related by a model of the form

$$d_{ij} = h_{ij} + e_{ij} \ (i, j = 1, ..., n), \tag{4.23}$$

where $e_{ij}$ is an 'error' term. This section describes several more
general models for the dissimilarities that have been proposed by J.
D. Carroll and collaborators. Some additional models are described
by De Soete and Carroll (1996).

The multiple trees model of Carroll and Pruzansky (1975, 1980)
assumes that the dissimilarities arise from an additive combination
of $t$ different valued trees:

$$d_{ij} = \Sigma_{r=1}^t h_{ijr} + e_{ij} \ (i, j = 1, ..., n), \qquad (4.24)$$

where $h_{ijr}$ denotes the ultrametric distance between the $i$th and
$j$th objects in the $r$th tree $(r = 1, ..., t)$. They illustrate this model
by applying it to several data sets, including the kinship terms
data, with the measure of dissimilarity between two terms being
given by the proportion of times that they were sorted into different
categories. Most of the subjects grouped together equivalent terms
describing individuals of different gender (e.g. 'brother' is perceived
as similar to 'sister') but some subjects categorized the terms solely
by gender. This distinction is clearly reflected in the results of
fitting the two-tree model shown in Fig. 4.16; the first tree also
distinguishes the nuclear family from the extended family, and the
gender-ambiguous term 'cousin' differs markedly from the other
terms in the second tree.

Descriptions follow of two generalizations of the multiple trees
model. First, Carroll and Pruzansky (1975, 1980) proposed a hybrid model:

$$d_{ij} = \Sigma_{r=1}^t h_{ijr} + g_{ij} + e_{ij} \ (i, j = 1, ..., n), \qquad (4.25)$$

where $g_{ij}$ denotes the difference between the $i$th and $j$th objects
in some other structure, e.g. the squared distance between them
in some Euclidean space. Secondly, for the case in which each of
$s$ different individuals provides a separate matrix of the pairwise
dissimilarities within the same set of $n$ objects, Carroll, Clark and
DeSarbo (1984) extend equation (4.24) by including an additional
subscript:

$$d_{ijk} = \Sigma_{r=1}^t h_{ijkr} + e_{ijk} \ (i, j = 1, ..., n; k = 1, ..., s), \qquad (4.26)$$

where $h_{ijkr}$ denotes the ultrametric distance between the $i$th and
$j$th objects in the $r$th tree associated with the $k$th individual, and
the constraint is imposed that the $r$th tree has the same topology

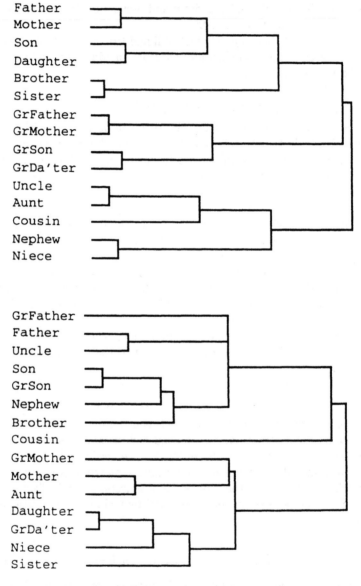

Figure 4.16 *Two-tree representation of the kinship terms data. Figure redrawn from Carroll and Pruzansky (1980).*

(but possibly different node heights) for each of the $s$ individuals $(r = 1, ..., t)$.

Given more elaborate models of this kind, it is clearly important to be able to specify appropriate models, e. g. to select a relevant value of $t$ in (4.24) and decide if inclusion of $g_{ij}$ in (4.25) markedly improves the fit of the model. To date, the assessment of the adequacy of such models has largely been based on the interpretability of the results, and there remains a need for more formal methods of assessment.

# CHAPTER 5

# Other clustering procedures

## 5.1 Fuzzy clustering

A partition of a set of objects into a specified number of classes may be a considerable oversimplification of the structure in the data set: in particular, it may be that there are some objects which should definitely be assigned to certain classes but other objects whose class membership is much less obvious. Fuzzy clustering is one approach which provides fuller details about the structure within each class; such information can also be provided by graphical representations and summary statistics for clusters, as described in Chapters 6 and 7.

As an illustration, Fig. 5.1 depicts a set of objects described by two quantitative variables. The dashed line indicates the partition of these data into two classes that minimizes the sum of squares criterion, $P(H1; \Sigma)$. However, it is clear that the objects labelled $A - D$ are less closely associated with their respective classes than other objects. Fuzzy clustering aims to represent these relationships by associating with each object a set of *membership functions*

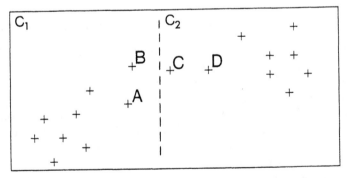

Figure 5.1 *A partition of a set of eighteen objects into two classes, with transitional objects labelled A, B, C, D.*

which specify the strength with which each object is regarded as belonging to each class: thus, $u_{ir}$ denotes the membership function of the $i$th object for the $r$th class, with high values of $u_{ir}$ denoting a strong degree of membership. The set of values $\mathbf{u} \equiv (u_{ir})$ is required to satisfy

$$0 \leq u_{ir} \leq 1 \ (i = 1, ..., n; r = 1, ..., c) \tag{5.1}$$

$$0 < \Sigma_{i=1}^{n} u_{ir} < n \ (r = 1, ..., c) \tag{5.2}$$

$$\Sigma_{r=1}^{c} u_{ir} = 1 \ (i = 1, ..., n). \tag{5.3}$$

If condition (5.3) is omitted, $(u_{ir})$ are referred to as 'possibilities', whose sum over the classes need not be 1 for each object.

Early work in fuzzy clustering (Ruspini, 1969, 1970) had fairly strong links with aspects of multivariate mixture models: for example, a probability density function was assumed known and the 'degree of belongingness' of the vector $\mathbf{x_i}$ to the class $S_r$ was defined to be the posterior probability $P(S_r|\mathbf{x_i})$. However, an intuitively-defined criterion was optimized, with no reference being made to relevant areas of statistical methodology such as likelihood theory.

More recent work has moved farther away from established statistical methodology. For example, a fuzzy analogue of the sum of squares criterion, $P(H1, \Sigma)$, is (Bezdek, 1974; Dunn, 1974)

$$F_{cm}(\mathbf{u}) \equiv \Sigma_{r=1}^{c} \Sigma_{i=1}^{n} u_{ir}^{m} D_{ir}, \tag{5.4}$$

where

$$D_{ir} \equiv \Sigma_{k=1}^{p} (x_{ik} - v_{rk})^{2} \ (i = 1, ..., n; r = 1, ..., c), \tag{5.5}$$

$\{v_{rk}(k = 1, ..., p)\}$ are the coordinate values of the 'centre' of the $r$th class $(r = 1, ..., c)$, and $m \ (\geq 1)$ is a parameter controlling the degree of fuzziness of the solution. The values of $(u_{ir})$ which minimize $F_{cm}(\mathbf{u})$ subject to conditions (5.1) − (5.3) are sought. Minimization of $F_{cm}(\mathbf{u})$ when $m = 1$ leads to a 'hard partition' solution in which each $u_{ir}$ is either 0 or 1 (Fisher, 1958), whereas each $u_{ir}$ tends to $1/c$ as $m \to \infty$. Many investigators have carried out analyses using $m = 2$.

Values $\mathbf{u}$ which minimize $F_{cm}(\mathbf{u})$ $(1 < m < \infty)$ can be sought using a two-stage procedure of successively updating $(v_{rk})$ and $(u_{ir})$, using the formulae

$$v_{rk} = \Sigma_{i=1}^{n} u_{ir}^{m} x_{ik} / \Sigma_{i=1}^{n} u_{ir}^{m} \ (r = 1, ..., c; k = 1, ..., p) \qquad (5.6)$$

and, if $D_{ir} > 0 \ (r = 1, ..., c)$,

$$u_{ir} = \left[ \Sigma_{s=1}^{c} (D_{ir}/D_{is})^{1/(m-1)} \right]^{-1} \ (i = 1, ..., n; r = 1, ..., c). \quad (5.7)$$

If at any time during the iterations $D_{ir} = 0$ for $r \in R_i, u_{ir}$ is assigned the value 0 for all $r \notin R_i$ and is given an arbitrary value satisfying conditions (5.1) $-$ (5.3) for all $r \in R_i$.

This two-stage updating procedure is repeated until the changes in **u** between successive iterations are smaller than some user-specified threshold value. Convergence results for this algorithm are summarized by Hathaway and Bezdek (1988).

When this methodology was applied to the data set depicted in Fig. 5.1, the membership functions of the four labelled objects were found to be

$A : (0.88, 0.12); \ B : (0.75, 0.25); \ C : (0.47, 0.53); \ D : (0.16, 0.84).$

All other objects in Fig. 5.1 have a membership function for 'their' class whose value is at least 0.96.

Various measures of the 'fuzziness' of a solution have been proposed, often with a view to identifying appropriate values of $c$ (e.g. Roubens, 1978; Windham, 1982; Xie and Beni, 1991). Some theoretical and simulation investigations of several measures are reported by Pal and Bezdek (1995), but there remains a need for more detailed study.

The form of Equation (5.4) indicates that other fuzzy clustering models are obtainable by replacing $D_{ir}$ by some other measure of distance between object and cluster centre, such as the Mahalanobis distance (Bezdek, 1974), city block metric (Trauwaert, 1987; Miyamoto and Agusta, 1995) or $\infty$-norm (Bobrowski and Bezdek, 1991). Other fuzzy clustering criteria are reviewed by Bezdek (1981, 1987).

However, just as with traditional classification criteria, a fuzzy clustering criterion implicitly specifies an underlying model for data, and can misrepresent the structure present in a data set. This is illustrated in Fig. 5.2, which summarizes the results of a fuzzy sum of squares clustering of Diday and Govaert's data into $c = 3$ classes. A 'hard' partition has been obtained from the fuzzy

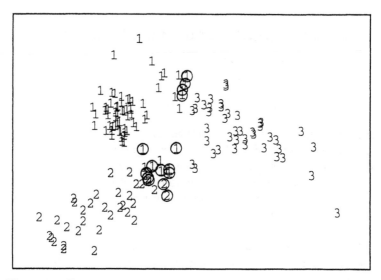

Figure 5.2 *Partition of Diday and Govaert's data into three classes obtained by 'hardening' the fuzzy sum of squares solution; objects whose largest membership function is less than 0.5 are circled.*

solution by assigning each object to the class for which its membership function is highest; objects whose largest membership function is smaller than 0.5 are shown circled. By comparing Fig. 5.2 with Figs. 3.4 and 3.5, it can be seen that fuzzy sum of squares clustering has 'misclassified' a few more objects than the sum of squares criterion $P(H1, \Sigma)$ and has the same tendency to find spherical-shaped classes. Theoretical reasons for expecting this feature are presented by Trauwaert, Kaufman and Rousseeuw (1991), who carried out an investigation along the lines of the ones described in Section 3.6. In Expressions like (3.15) for the likelihood under a normal components model, they replaced

$$\Sigma_{r=1}^{c}\Sigma_{i \in E_r} f(i,r) \text{ by } \Sigma_{r=1}^{c}\Sigma_{i=1}^{n} u_{ir} f(i,r),$$

where $f(i,r)$ denotes a function of $i$ and $r$, and $E_r$ denotes the set of objects belonging to the $r$th component. Trauwaert, Kaufman and Rousseeuw (1991) consider the effect of replacing $u_{ir}$ by $u_{ir}^{m}$, allowing $m$ to take different values in different parts of the 'likelihood' function, and making various assumptions about the covariance matrix of the $r$th class, $\Sigma_r$, e. g.

$$\Sigma_r = \Sigma \text{ and } \Sigma_r = \sigma^2 I \ (r = 1, ..., c).$$

Maximizing the 'likelihood' under these two assumptions leads to minimization of, respectively, a fuzzy version of Friedman and Rubin's (1967) det $W$ criterion and the fuzzy sum of squares criterion $F_{cm}(\mathbf{u})$ (5.4, 5.5). As stressed earlier, clustering criteria are not 'model-free', and due consideration needs to be given to the choice of relevant methods of analysis for data.

## 5.2 Constrained classification

In some classification problems, there are reasons for restricting the set of allowable solutions. Such restrictions commonly occur when the objects to be classified have a spatial context. For example, given the properties of soil profiles at many different sites, it is convenient for the purposes of soil management to create 'parcels' of land within which the variation in soil properties is small (Webster and Burrough, 1972); thus, a parcel of land comprises a set of spatially-contiguous sites (or objects) which have similar properties, and should be reasonably large and compact. More recently, remote sensing techniques have provided information about large numbers of pixels, which it can be useful to summarize using constrained classification methodology. Other examples occur in the field of marketing: in defining sales territories, it can be relevant to impose constraints on the membership or properties of classes (DeSarbo and Mahajan, 1984).

In the classification of spatial data, the dissimilarity between each pair of objects can be defined to include not only a measure of their physical differences but also a term depending on the distance that separates them, and these dissimilarities can be analysed using standard clustering procedures (Webster and Burrough, 1972; Perruchet, 1983). If the 'distance' term is weighted sufficiently heavily, partitions into classes of spatially-contiguous objects will be obtained, but it can be difficult to specify the weights in an objective manner.

A more satisfactory approach requires the definition of a contiguity matrix $\mathbf{C} \equiv (c_{ij})$, where $c_{ij} = 1$ if the $i$th and $j$th objects are regarded as contiguous, and 0 if they are not $(i, j = 1, ..., n)$. In the corresponding contiguity graph, each object is represented by a vertex of the graph and an edge links the $i$th and $j$th vertices if $c_{ij} = 1$. In the following description, it is assumed that the contiguity graph is connected; if this is not the case, each connected component of the graph would be analysed separately.

The specification of **C** is reasonably straightforward if the objects comprise areas of land, two areas being regarded as contiguous if they have a common boundary, although it is necessary to decide whether or not two areas are neighbours if they meet only at a single point; thus, areas arranged in a rectangular grid may have four or eight neighbours. If objects have no spatial extent, but are located at irregular positions in space, the following constructs from the field of computational geometry (Preparata and Shamos, 1988) can assist the construction of contiguity graphs; they were originally defined for two-dimensional point patterns, but have been extended to higher dimensions.

1. The Voronoi diagram associates with each point the region of space that is closer to it than to any other point. In its dual, the Delaunay triangulation (DT), an edge joins each pair of points whose regions have a common boundary. Attention is often restricted to relationships within a 'window' containing all the points; this will modify the Delaunay triangulation if some regions are neighbours only outside the window. Toussaint (1980a) and Aurenhammer (1991) discuss properties of these constructs.

2. In the Gabriel graph (GG) (Gabriel and Sokal, 1969; Matula and Sokal, 1980), an edge joins the $i$th and $j$th points if and only if all other points lie outside the circle whose diameter is that edge.

3. In the relative neighbourhood graph (RNG) (Lankford, 1969; Toussaint, 1980a, b), an edge joins the $i$th and $j$th points if and only if the distance between them is no greater than the distance of either of them from any other point.

A fourth graph is given by the minimum spanning tree (MST), described in Section 3.4.1. These four graphs are nested, as follows (Howe, 1979; Matula and Sokal, 1980; Toussaint, 1980a, b):

$$MST \subseteq RNG \subseteq GG \subseteq DT.$$

For example, Fig. 5.3 depicts the Voronoi diagram of a set of points located within a rectangular region of the plane, and Fig. 5.4 shows the contiguity graphs of these points provided by the four constructs given above.

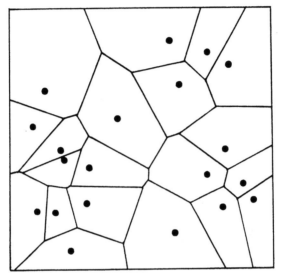

Figure 5.3  *Voronoi diagram of a set of points in the plane. Modified and redrawn from Toussaint (1980a).*

These constructs from computational geometry can assist in the specification of contiguity graphs that are relevant for constrained classification studies, but investigators should be prepared to modify such graphs if they consider that this would be appropriate for their particular data sets; for example, the earth is not a featureless plane and the presence of mountain chains may indicate that some edges should not be included because the objects that they link should not be regarded as neighbours. A detailed analysis would make use of more than one contiguity graph, and note for further consideration single edges whose removal would alter the results of a classification.

Given relevant contiguity graphs, constrained partitions and hierarchical classifications can be obtained using modified versions of clustering algorithms described in Chapters 3 and 4. Thus, agglomerative algorithms can be constrained so as to ensure that only contiguous classes of objects are amalgamated at each stage (e.g. Berry, 1968; Spence, 1968; Webster and Burrough, 1972); additional constraints may be imposed to limit the size of any class (Lebart, 1978). Two classes are usually regarded as contiguous if there are two objects, one from each class, which are linked by an

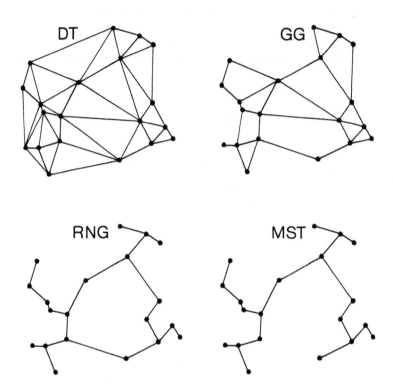

Figure 5.4 *Contiguity graphs for the set of points shown in Fig. 5.3: DT = Delaunay triangulation; GG = Gabriel graph; RNG = relative neighbourhood graph; MST = minimum spanning tree. Modified and redrawn from Toussaint (1980a).*

edge in the contiguity graph. However, this can lead to reversals in the 'hierarchical classification', i.e. one can obtain classes $C_i$ and $C_j$ for which $C_i \subset C_j$ but their heights $h$ are such that $h(C_i) > h(C_j)$. Ferligoj and Batagelj (1982) prove that necessary and sufficient conditions to guarantee an absence of reversals in the results of a constrained version of Lance and Williams's (1966b, 1967) general agglomerative algorithm (Equation (4.7) with $\delta_i = \delta_j = \epsilon = 0$) are:

$$\gamma \geq -\min(\alpha_i, \alpha_j) \qquad (5.8)$$

$$\alpha_i + \alpha_j \geq 0 \qquad (5.9)$$

$$\min(\alpha_i + \alpha_j, \gamma + \min(\alpha_i, \alpha_j)) + \beta \geq 1. \qquad (5.10)$$

Constrained versions of iterative relocation algorithms (Ferligoj and Batagelj, 1982) allow an object to be moved to a different class only if it is contiguous to at least one member of that class and if the relocation of the object would not cause its former class to become disconnected. Other algorithms include some based on extracting a single cluster at a time (Gabriel and Sokal, 1969; Lefkovitch, 1980), replacing contiguity graphs by spanning trees (Maravalle and Simeone, 1995) and hybrid split-and-merge algorithms (Horowitz and Pavlidis, 1976; Fukada, 1980). Boundary or edge detection algorithms from the field of image segmentation (Peli and Malah, 1982; Huang and Tseng, 1988) may also be relevant in the construction of some constrained classifications.

A special case of constrained classification occurs when there is an ordering on the set of objects, specified for example by time, stratigraphy or distance along a transect, and the classification is required to respect the ordering. In this case, class boundaries can be indicated by inserting 'markers' between pairs of neighbouring objects. With the objects numbered 1, 2, ..., $n$ in the order specified by the constraint, let a marker labelled $m$ denote the presence of a class boundary between objects $m$ and $(m + 1)$. The objects can be partitioned into $c$ classes of contiguous objects by specifying $(c - 1)$ markers taking distinct values from $\{1, 2, ..., n - 1\}$. If the heterogeneity of a partition is defined to be the sum of the within-class heterogeneities, optimal solutions can be obtained using dynamic programming techniques (Fisher, 1958; Hawkins and Merriam, 1973). Let $H(i, j)$ denote the heterogeneity of the class comprising objects $\{i, i + 1, ..., j\}$, and $T(c, k)$ denote the heterogeneity of the optimal partition into $c$ classes of objects $\{1, 2, ..., k\}$. Then, the values $T(c, k)$ $(k = 1, ..., n; c = 1, ..., n)$ are constructed recursively as follows:

$$T(1, k) = H(1, k) \quad (k = 1, 2, ..., n) \qquad (5.11)$$

and, for $k = c, c + 1, ..., n; c = 2, ..., n$,

$$T(c, k) = \min_{(c - 1 \leq j \leq k - 1)} \{T(c - 1, j) + H(j + 1, k)\}. \qquad (5.12)$$

The (not necessarily unique) set of markers specifying an optimal partition into $c$ classes can be obtained by a trace-back proce-

Table 5.1 *Constrained partitions of the Abernethy Forest 1974 data into c classes for values of c between 2 and 8; the markers indicate the last members of the first c − 1 classes.*

| c | Markers | | | | | | |
|---|---|---|---|---|---|---|---|
| 2 | 15 | | | | | | |
| 3 | 15 | 32 | | | | | |
| 4 | 15 | 33 | 41 | | | | |
| 5 | 15 | 33 | 41 | 45 | | | |
| 6 | 15 | 32 | 34 | 41 | 45 | | |
| 7 | 15 | 26 | 32 | 34 | 41 | 45 | |
| 8 | 8 | 15 | 26 | 32 | 34 | 41 | 45 |

dure: thus, the value of $j$ ($j^*$, say) which defines $T(c, n)$ in Equation (5.12) gives the value of the last marker, and the second last marker can be identified from the definition of $T(c - 1, j^*)$, etc. For many definitions of the heterogeneity of a class, optimal partitions for different values of $c$ need not be hierarchically nested. Order-constrained hierarchical classifications resulting from several different clustering criteria have also been obtained using agglomerative (Gordon and Birks, 1972; Mehringer, Arno and Petersen, 1977), divisive (Gill, 1970; Gordon and Birks, 1972), and direct optimization (Gordon, 1973) algorithms.

Constrained classifications of the Abernethy Forest 1974 data were obtained by minimizing the sum of squares criterion $P(H1, \Sigma)$ subject to the constraint that all classes comprise consecutively-numbered samples. Table 5.1 presents the markers defining the optimal sum of squares constrained partitions into $c$ classes for all values of $c$ between 2 and 8, from which it is seen that the partitions are not completely hierarchically-nested. A divisive algorithm, based on minimizing the sum of squares by splitting an existing class into two at each stage, provides results which differ from those in Table 5.1 only in the partitions into four and five classes. The results of these analyses are very similar to the unconstrained sum of squares partition of these data into five classes that is reported in Section 3.2.3, differing only in the treatment of samples 33 and 34.

Although contiguity in space or time provides the most common constraint in classification studies, other types of restriction can be relevant. Constraints can be imposed on the membership, size or properties of classes by specifying conditions on the class membership matrix $\mathbf{M} = (m_{ir})$, where $m_{ir} = 1$ (resp., 0) if the $i$th object belongs (resp., does not belong) to the $r$th class ($i = 1, ..., n; r = 1, ..., c$) (Mahajan and Jain, 1978; Klastorin and Watts, 1981; DeSarbo and Mahajan, 1984). For example,

- the $i$th and $j$th objects can be forced to belong to different classes by imposing the condition

$$m_{ir} + m_{jr} \leq 1 \ (r = 1, ..., c) \qquad (5.13)$$

- the $r$th class can be required to contain between $a_r$ and $b_r$ objects by specifying

$$a_r \leq \Sigma_{i=1}^{n} m_{ir} \leq b_r \qquad (5.14)$$

- the average within-class values of a variable $V$ in the $r$th and $s$th classes can be required to differ by no more than $\epsilon$ by specifying

$$\left| \frac{\Sigma_{i=1}^{n} m_{ir} V_i}{\Sigma_{i=1}^{n} m_{ir}} - \frac{\Sigma_{i=1}^{n} m_{is} V_i}{\Sigma_{i=1}^{n} m_{is}} \right| \leq \epsilon \qquad (5.15)$$

Such constraints can be included in a penalty function which is added to the objective function to be optimized. Constraints like (5.15) may prove impossible to satisfy.

The parsimonious trees described in Section 4.2.4 and 'individual differences' clustering models defined in Equations (4.26) and (5.17) can also be regarded as examples of constrained classifications.

More detailed surveys of work in constrained classification are presented by Murtagh (1985) and Gordon (1996b).

## 5.3 Overlapping classification

Classes of objects belonging to a partition are disjoint from one another, and classes in the hierarchical classifications described in Chapter 4 are either disjoint or nested. On occasion, it is relevant to obtain a classification in which different classes can have some, but not necessarily all, of their objects in common. Thus, in Example 4 in Chapter 1, individuals in a social network could have several different aims or attributes, some but not all of which are shared with other individuals. Some of the methodology for identifying single classes described in Section 3.4.3 can provide overlapping classifications. This section describes several different methodological approaches for representing such relationships.

### 5.3.1 Additive clustering

In the additive clustering model (Shepard and Arabie, 1979; Arabie and Carroll, 1980), each class has an associated weight, and the similarity $s_{ij}$ between the $i$th and $j$th objects is regarded as being composed of a sum of the weights of the classes in which they occur together:

$$s_{ij} \cong \Sigma_{r=1}^{c-1} w_r m_{ir} m_{jr} + a \ (i, j = 1, ..., n), \qquad (5.16)$$

where $w_r$ is the weight of the $r$th class, $m_{ir} = 1$ (resp., 0) if the $i$th object belongs (resp., does not belong) to the $r$th class ($i = 1, ..., n; r = 1, ..., c-1$), and $a$ is an additive constant (or the weight of a $c$th class to which all the objects belong).

Arabie and Carroll (1980) fit this model using an alternating least squares algorithm that incorporates a penalty function intended to force each $m_{ir}$ to be either 0 or 1. They note the possibility of obtaining different solutions with similar values of the goodness-of-fit criterion. The number of classes, $c$, has to be specified by the user, and Arabie and Carroll (1989) advocate obtaining solutions for a range of values of $c$, and selecting a value that balances goodness-of-fit and interpretability.

Table 5.2 presents results reported by Arabie and Carroll (1980) of fitting the additive clustering model with $c = 9$ to the acoustic confusion data; the additive constant $a$ takes the value 0.049. Also presented in Table 5.2 are Arabie and Carroll's (1980) interpretations of each of the classes, with interpretations which they regarded as questionable being shown in parentheses. The results are consistent with the earlier analysis of these data portrayed in Figs. 4.5 and 4.7.

More general formulations than the additive clustering model have been proposed. Mirkin (1987) describes qualitative factor analysis methodology, in which the aim is to approximate a similarity matrix in models which include 0/1 matrices with simple structure. Carroll and Arabie (1983) describe an *individual differences clustering* model, in which each of a set of $t$ observers

Table 5.2 *Analysis of the acoustic confusion data reported by Arabie and Carroll (1980).*

| Weights | Members of Class | Interpretation |
|---------|------------------|----------------|
| 0.814 | $fa, thin$ | front unvoiced fricatives |
| 0.729 | $va, that$ | front voiced fricatives |
| 0.577 | $da, ga$ | back voiced stops |
| 0.487 | $pa, ka, ta$ | unvoiced stops |
| 0.428 | $ba, va$ | (front voiced consonants) |
| 0.348 | $pa, ka$ | unvoiced stops, omitting $ta$ |
| 0.162 | $ba, da, ga,$ $that, za, zha$ | voiced consonants, omitting $va$ |
| 0.116 | $pa, ka, fa,$ $thin, sa, sha$ | (unvoiced consonants, omitting $ta$) |

provides a similarity matrix describing the same set of $n$ objects, and it is assumed that the objects belong to a common (possibly overlapping) set of classes which might be weighted differently by different observers, i.e.

$$s_{ijl} \cong \Sigma_{r=1}^{c-1} w_{lr} m_{ir} m_{jr} + a_l \ (i, j = 1, ..., n; l = 1, ..., t), \quad (5.17)$$

where $s_{ijl}$ is the similarity between the $i$th and $j$th objects provided by the $l$th observer, $m_{ir}$ is defined as before, $w_{lr}$ is the weight that the $l$th observer assigns to the $r$th class, and $a_l$ is an additive constant for the $l$th observer $(i, j = 1, ..., n; l = 1, ..., t; r = 1, ..., c-1)$.

### 5.3.2 $B_k$ methods

An alternative approach to obtaining overlapping classifications is provided by the $B_k$ $(k = 1, 2, 3, ...)$ sequence of clustering methods (Jardine and Sibson, 1968). In the $B_k$ method, a maximum of $(k - 1)$ objects may belong to the overlap between any pair of classes. The $B_1$ method is the single link method.

Consider the graph theoretical representation described in Chapter 3, in which each object is represented by a vertex of a graph and an edge links the $i$th and $j$th vertices at level $h$ if the dissimilarity

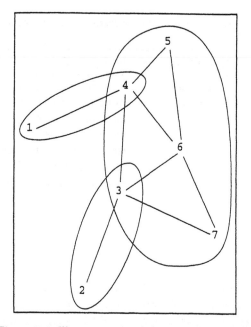

Figure 5.5 *Illustrating the definition of $B_2$ classes.*

between the corresponding pair of objects, $d_{ij}$, is no greater than $h$. The $B_k$ classes at level $h$ can be obtained as follows:

1. Each maximal complete subgraph of the graph defines an initial class

2. If two classes have more than $(k-1)$ objects in common, they are amalgamated

The classes resulting after all amalgamations have been carried out are $B_k$ classes.

As an illustration, Fig. 5.5 shows the graph at height $h = 305$ for the set of objects whose dissimilarity matrix is given in Table 3.2. The maximal complete subgraphs of this graph are $\{1, 4\}$, $\{2, 3\}$, $\{3, 4, 6\}$, $\{3, 6, 7\}$ and $\{4, 5, 6\}$. If $B_2$ classes are sought, the latter three classes do not remain separate because class $\{3, 4, 6\}$ has two objects in common with each of the other two. Hence, the $B_2$ classes at height 305 are $\{1, 4\}$, $\{2, 3\}$ and $\{3, 4, 5, 6, 7\}$, as shown in Fig. 5.5, each of the first two classes having one object in common with the class $\{3 - 7\}$.

Table 5.3 *Transformed dissimilarity matrix* $(d_{ij}^{(2)})$, *summarizing the* $B_2$ *analysis of the matrix of dissimilarities given in Table 3.2.*

| Object | | | | | | |
|--------|------|------|------|------|------|------|
| 2 | 565 | | | | | |
| 3 | 325 | 292 | | | | |
| 4 | 305 | 565 | 290 | | | |
| 5 | 325 | 565 | 293 | 98 | | |
| 6 | 325 | 565 | 149 | 181 | 293 | |
| 7 | 325 | 565 | 305 | 305 | 305 | 232 |
| Object | 1 | 2 | 3 | 4 | 5 | 6 |

Fig. 5.5 shows the $B_2$ classes for a single value of the height, $h$ = 305. By varying the value of $h$, the complete set of $B_2$ classes can be obtained. These can be summarized in a table of transformed dissimilarities $(\hat{d}_{ij}^{(2)})$, where $\hat{d}_{ij}^{(2)}$ denotes the lowest value of the height for which the $i$th and $j$th objects belong to the same $B_2$ class. Table 5.3 shows the values of these transformed dissimilarities resulting from a complete $B_2$ analysis of the set of seven objects. Table 4.3 shows a similar set of transformed dissimilarities resulting from a $B_1$ or single link analysis of these data. A comparison of these Tables with the original matrix of dissimilarities, shown in Table 3.2, illustrates the fact that the distortion imposed on the original matrix of dissimilarities is a non-increasing function of the amount of overlap; in fact $\hat{d}_{ij}^{(n-1)} = d_{ij}$ $(i, j = 1, ..., n)$. The set of transformed dissimilarities $(\hat{d}_{ij}^{(k)})$ can be represented graphically in so-called $k$-dendrograms, but these can contain crossings (i.e. may not be planar graphs) and become increasingly difficult to interpret for larger values of $k$. The algorithms that have been proposed for obtaining $B_k$ classifications (Jardine and Sibson, 1968; Cole and Wishart, 1970; Rohlf, 1975a) also make heavy demands on computing resources, and the methodology seems likely to be of use only in the analysis of small data sets.

### 5.3.3 Pyramids

A pyramid (Diday, 1984, 1986) specifies an ordering of the set of objects. Each of its constituent classes comprises an interval of this order, and two classes may have some but not all of their objects in common. Formally, a pyramid is a set $P$ of subsets of the set of objects $\Omega$ satisfying the following conditions:

$(i)$ $\Omega \in P$; $(ii)$ $\emptyset \notin P$; $(iii)$ $\{i\} \in P$ for all $i \in \Omega$;
$(iv)$ if $A, B \in P$, then $A \cap B \in P \cup \{\emptyset\}$;
$(v)$ there is an order on $\Omega$ such that $P$ is a set of intervals of this order.

Pyramids are generalizations of the hierarchical classifications described in Chapter 4, as can be seen from the comparison of the above definition with the definition of an $n$-tree given in Section 4.1. For example, if the set of six objects $\Omega \equiv \{1, 2, 3, 4, 5, 6\}$ is ordered in the natural order, the set of subsets $P$ comprising $\Omega$, the singleton subsets $\{i\}(i = 1,...,6)$, $\{1, 2\}, \{2, 3\}, \{5, 6\}, \{1-3\}, \{2-4\}$, and $\{1 - 4\}$ is a pyramid but not an $n$-tree. This pyramid is portrayed in the planar graph shown in Fig. 5.6, in which each internal node defines the class of objects that it subtends. No internal node can have more than two 'parent' or 'predecessor' nodes (Diday, 1984), hence the maximum number of internal nodes in a pyramid based on $n$ objects is $n(n - 1)/2$; by contrast, the trees considered in Chapter 4 have a maximum of $(n - 1)$ internal nodes.

As was done with rooted trees, a height $h$ can be associated with each of the subsets in a pyramid, to obtain an *indexed pyramid*, for which:

1. $h(A) = 0 \Leftrightarrow A$ is a singleton subset;

2. for all $A, B \in P, A \subset B \Rightarrow h(A) \leq h(B)$,

where $\subset$ denotes strict inclusion. The second condition has led to two different definitions of an indexed pyramid, which has been defined to be:

- *weakly indexed* if $A \subset B$ and $h(A) = h(B)$ implies that $A$ has two distinct predecessors

- *strictly indexed* if $A \subset B \Rightarrow h(A) < h(B)$.

By sectioning a pyramid at a specified value of the height, $h$, an *ordered* overlapping classification is obtained; for example, if the pyramid shown in Fig. 5.6 is sectioned at the height indicated by the arrow, the overlapping classification into the ordered classes $\{1 - 3\}, \{2 - 4\}, \{5, 6\}$ results.

There is a one-one correspondence between weakly indexed pyramids and Robinson dissimilarity matrices, i.e. matrices whose values are monotonically non-decreasing as one moves away from the

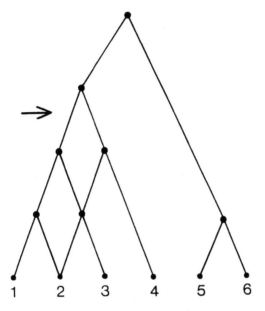

Figure 5.6  *A pyramid.*

diagonal within each row and within each column (Diday, 1986; Bertrand, 1995). Corresponding results for strictly-indexed pyramids are described by Durand and Fichet (1988) and Mirkin (1996, Section 7.5.1).

A pyramid can be constructed from a dissimilarity matrix using an agglomerative algorithm (Diday, 1986; Diday and Bertrand, 1986), or by minimizing a sum of squares measure of discordance which has been augmented by a penalty function (Gaul and Schader, 1994), the latter approach allowing treatment of missing values. Pyramids have not yet found widespread use in the analysis of data. Some examples given by Diday (1986) illustrate the fact that they can be difficult to interpret, although this difficulty may be at least partly caused by investigators' unfamiliarity with pyramids. Interpretation can be assisted by the suppression of some internal nodes, in an analogous fashion to the local parsimonious trees described in Section 4.2.4, but there remains a need for further investigations of the usefulness of pyramids in the analysis of data.

## 5.4 Conceptual clustering

In conceptual clustering methodology, an explicit definition of properties that are (broadly) satisfied by members of each class is derived during the course of obtaining a classification. The first two subsections describe two methods of classifying binary data, in which it is assumed that the objects are described by the same $p$ variables, each of which comprises the states '+' and '−'. Although not originally described in such terms, these two methods of analysis can be seen as natural precursors of conceptual clustering methods. The final subsection describes some more recent work, involving the classification of objects described by other types of variables.

### 5.4.1 Maximal predictive classification

As a motivation for his method of maximal predictive classification, Gower (1974) quoted a statement of Gilmour (1937) that

> 'a system of classification is the more natural the more propositions there are that can be made regarding its constituent classes.'

Each of the $c$ classes in a partition is assigned a *class predictor*, which is a dummy object defined by the most common state (in the class) for each variable; clearly, class predictors need not be uniquely defined if the class contains an even number of objects. This is illustrated by the data set shown in Table 5.4, in which ten objects are described by five binary variables and a partition into three classes is considered. For example, the class predictor of class $C_1$, comprising objects $\{1 - 4\}$, is $(+, +, -, +, +)$. This class predictor correctly predicts the states of all variables for object 1, and the states of four out of the five variables for each of the other three objects in class $C_1$. Let $n_{rs}$ denote the number of correct predictions that would be made when the $r$th class predictor is used to predict the properties of the objects in the $s$th class ($r, s = 1, ..., c$). The total number of correct predictions made by the class predictors for members of their own classes is

$$W_c \equiv \Sigma_{r=1}^c n_{rr}, \qquad (5.18)$$

and the average number of correct predictions which would be made by comparing each object with a predictor belonging to a different class is

Table 5.4 *Illustrating the method of maximal predictive classification.*

| Class $C_s$ | Object | Variable | | | | | Class predictor | Number of correct predictions made by class's own predictor | Total number of correct predictions, $n_{rs}$, made by predictor of class $C_r$ | | |
|---|---|---|---|---|---|---|---|---|---|---|---|
| | | 1 | 2 | 3 | 4 | 5 | | | $r=1$ | $r=2$ | $r=3$ |
| $C_1$ | 1 | + | + | − | + | + | | 5 | | | |
| | 2 | + | + | − | − | + | + + − + + | 4 | 17 | 7 | 11 |
| | 3 | + | − | − | + | + | | 4 | | | |
| | 4 | − | + | − | + | + | | 4 | | | |
| $C_2$ | 5 | + | + | + | + | − | | 4 | | | |
| | 6 | + | + | + | − | − | + + + − − | 5 | 6 | 13 | 3 |
| | 7 | + | − | + | − | − | | 4 | | | |
| $C_3$ | 8 | − | − | − | − | + | | 5 | | | |
| | 9 | − | + | − | − | + | − − − − + | 4 | 8 | 3 | 13 |
| | 10 | − | − | − | + | + | | 4 | | | |

$$B_c \equiv \Sigma_{r=1}^{c} \Sigma_{s=1(s \neq r)}^{c} n_{rs} / (c - 1). \tag{5.19}$$

For example, the partition into three classes shown in Table 5.4 has $W_3 = 43$ and $B_3 = 19$.

For each value of $c$, a partition which maximizes $W_c$ is sought, using an approximating iterative relocation algorithm. In order to select an appropriate number of classes, Gower (1974) advocates finding the value of $c$ for which $W_c - B_c$ is maximized. An application of the methodology to bacteriological data is presented by Barnett, Bascomb and Gower (1975).

### 5.4.2 Monothetic divisive algorithms

Divisive clustering algorithms successively split the data into a larger number of classes by dividing one of the existing classes, usually into two sub-classes; hence, partitions into different numbers of classes are hierarchically nested. Although there are efficient algorithms for carrying out the division so as to optimize some criteria (see Section 4.2.3), there are $2^{n-1} - 1$ different ways in which $n$ objects can be divided into two non-empty classes (Edwards and Cavalli-Sforza, 1965), and as the size of the data set increases it rapidly becomes computationally infeasible to examine all possible bi-partitions in order to identify an optimal one.

Monothetic divisive algorithms restrict attention to a much smaller set of bi-partitions by considering only divisions specified by the states of a single variable, hence the term 'monothetic': defined by a single species or variable. Thus, if division is based on the states of the $k$th variable, all of the objects under investigation which have the $k$th variable in the state '+' are placed in one sub-class, while objects with the $k$th variable in the state '−' are placed in another sub-class. The particular variable used for the division is chosen so as to ensure that the resulting sub-classes are as homogeneous as possible, hence the aim is to find the single variable which is the best predictor of the differences within the class of objects under study. At each stage of the algorithm, one of the existing classes is optimally divided into two in this manner, the process being repeated until an appropriate number of classes has been obtained.

Monothetic divisive algorithms appear to have been used most widely in the analysis of ecological quadrats (Example 2 in Chapter

1); in this case, an object is a small area of land which is described in terms of the presence (+) or absence (−) of certain species of plants. For example, Lance and Williams (1968) suggest measuring the variability of a class $A$, containing $n_A$ quadrats, as follows: if the $k$th species belongs to state '+' in $a_k$ of the $n_A$ quadrats, the 'information content' of class $A$ is defined by

$$I_A \equiv pn_A \log n_A - \Sigma^p_{k=1}[a_k \log a_k + (n_A - a_k)\log(n_A - a_k)]. \quad (5.20)$$

If $A$ is divided into the sub-classes $B$ and $C$, the 'change in information' is defined to be

$$\Delta I(A|B,C) \equiv I_A - I_B - I_C. \quad (5.21)$$

Lance and Williams (1968) advocate selecting at each stage the monothetic division that maximizes $\Delta I$; a range of other divisive criteria have also been proposed (e.g. Williams and Lambert, 1959; Crawford and Wishart, 1967).

This methodology is illustrated by application to the European fern data, in which 65 regions of Europe are described by the presence or absence of 144 species of fern. The first four divisions of the data using the information content criterion are as shown in Fig. 5.7: thus, the first division partitions the regions into two sub-classes, with species PS (*Polystichum setiferum*) being present in all regions in the first sub-class and absent from all regions in the second sub-class; the second division partitions the first of these sub-classes on the basis of the presence or absence of species LC (*Lycopodium clavatum*).

The geographical extent of each of the five classes is shown in Fig. 5.8, in which dashed lines delimit the area that was investigated and denote the boundaries between the classes. Each class is assigned a label, which is intended to provide an indication of the geographical area in which its regions are located. With few exceptions, the classes comprise sets of neighbouring regions; the island groups of the Azores and Svalbard, shown as insets, are assigned to classes SM and EC, respectively. Table 5.5 explains the class labels and provides a definition of the properties of each class in terms of the presence or absence of a small number of species of fern.

Monothetic divisive algorithms have some similarities with binary diagnostic keys or decision trees (Payne and Preece, 1980;

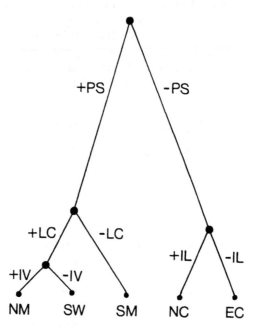

Figure 5.7 *The first four monothetic divisions of the European fern data.*
*The labels of the five classes and the four species specifying the divisions*
*are defined in Table 5.5.*

Table 5.5 *Description of five classes found when the European fern data*
*are analysed by a monothetic divisive algorithm using the information*
*content criterion.*

| Class label | Geographical extent of class | Species definition of class |
|---|---|---|
| SM | Southern Mediterranean | +PS −LC |
| NM | Northern Mediterranean | +PS +LC + IV |
| SW | Rest of south and west Europe | +PS +LC −IV |
| NC | North central Europe | −PS +IL |
| EC | East and central Europe | −PS −IL |

Species labels: IL = *Isoetes lacustris*; IV = *Isoetes velata*;
LC = *Lycopodium clavatum*; PS = *Polystichum setiferum*.

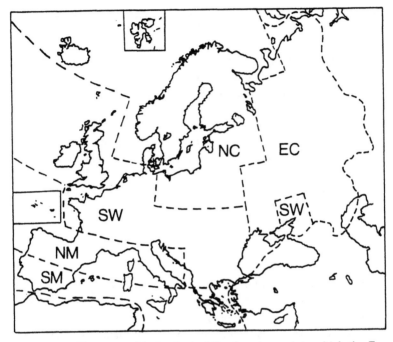

Figure 5.8 *The geographical extent of the five classes into which the European fern data are partitioned; the labels of the classes are given in Table 5.5.*

Breiman et al., 1984), but also important differences. For example, Fig. 5.7 may be regarded as a simple binary diagnostic key, to be used for the identification of a set of five 'entities', {NM, SW, SM, NC, EC}. If one were presented with a new object that was known to be identical to one of these five entities, one could rapidly identify it by carrying out a sequence of binary 'tests', starting with test PS. At each node that is encountered, one proceeds down the branch of the tree indicated by the result of the test, until a terminal node is reached. In the construction of diagnostic keys, each of the objects in a training set has an associated label, and one wishes to select a subset of the given set of tests which will allow a new object to be identified correctly with the least expected cost. If the costs of carrying out the tests do not differ markedly and each of the objects is equally likely to be presented for identification, efficient diagnostic keys tend to divide the class of objects under consideration at any stage into roughly equal-sized sub-classes. In the context of classification (as the word is used in this book), the (class) labels of the objects are not known, but require to be established. Further, there is no reason to suppose that divisions should be into sub-classes of roughly equal size, and inappropriate clustering criteria can give misleading summaries of the class structure present in data.

Classification based on monothetic division is sometimes regarded as too strict, in that an object can be more similar to objects belonging to other classes, differing from them only in one of the variables that happened to be selected for dividing the data. To counteract this effect, division is sometimes followed by some re-allocation of objects before the final classes are displayed (e.g. Crawford and Wishart, 1968).

### 5.4.3 More recent work

More recently, computer software has been developed to implement several different approaches to conceptual clustering. These generally define a measure of the adequacy of a classification and require iterative search strategies for its optimization. There can be a considerable number of heuristic elements in both the measure and the search strategy. As an illustration of the main ideas, descriptions are presented of the CLUSTER/2 (Michalski and Stepp, 1983) and COBWEB (Fisher, 1987, 1996) systems. Each of these systems is oriented towards the classification of objects described by categorical variables and requires continuous variables to be discretized.

CLUSTER/2's measure of classification adequacy is based on a combination of features, including the fit between data and classification, and the simplicity, accuracy and distinctness of the class descriptors. A set of $c$ seeds is specified, around which a partition into $c$ classes is to be constructed. Associated with each seed is a set of maximal *complexes* that cover (describe) it but no other seed. A complex is a logical product of variables, e.g. [height > 190 cm] & [eye colour = blue or green] & [70 kg < weight < 90 kg]. The set of complexes (called a *star*) associated with a seed will in general cover other objects in the data set. Each star is reduced and simplified as far as possible subject to its continuing to cover its class of objects. Objects covered by more than one star are removed and new stars are calculated for the reduced data set. A class is then optimally selected for each of the removed objects and new stars are calculated. New seeds are then selected, one from each class, and the process is repeated until no further improvements in the classification are obtained; the final stars provide a conceptual description of the classes. Hierarchical classifications can also be obtained.

COBWEB comprises an incremental algorithm for obtaining a hierarchical conceptual classification. Associated with each internal node in the tree diagram is a summary of the class of objects it subtends, in terms of the proportions of its constituent objects that belong to each state of each categorical variable. A measure of the quality of a class $C_r$ is defined:

$$Q(C_r) \equiv \text{prob}(C_r)[\Sigma_k \Sigma_s \text{prob}(V_{ks} \mid C_r)^2 - \Sigma_k \Sigma_s \text{prob}(V_{ks})^2],$$

where $V_{ks}$ denotes the event that the $k$th categorical variable belongs to its $s$th state. The quality of the partition into classes $\{C_1, ..., C_c\}$ can be measured by $\Sigma_{r=1}^c Q(C_r)/c$.

The hierarchical classification is recursively constructed by adding one object at a time. This object starts at the node comprising all objects and moves down the tree, with node descriptions being updated as it passes. Depending on the values of the quality measures defined above, the object can be assigned to an existing class, a new class can be created, or existing classes can be combined or subdivided. After a tree comprising the complete set of objects has been constructed, objects can be removed from the tree and optimally reclassified, in an attempt to obtain solutions that are less dependent on the order in which objects have been inserted into the tree. Some pruning of the lower branches of a tree can also be undertaken; methodology for doing this has also been developed in the context of simplifying decision trees (Breiman et al., 1984, Chapter 10; Quinlan, 1987). A conceptual description of each class in the classification is provided by the summary information recorded at each internal node of the tree diagram.

Discussions of some other conceptual clustering algorithms are provided by Fisher (1996) and Mirkin (1996, Section 3.2.5). There are also links between conceptual clustering and the classification of symbolic data, described in the next section.

The description of classes is an integral and important part of an investigation. However, it may be carried out during the course of obtaining a classification, as described in this section, or after the classification has been obtained; further discussion of the topic is postponed to Section 7.3.

## 5.5  Classification of symbolic data

For the types of data considered thus far, each entry in a pattern matrix has usually comprised a single value or category (cf. Section 2.2.5). The term 'symbolic data' refers to the description of individual objects, or classes of individual objects, by more general types of variable (Diday, 1988). Three examples of such variables are as follows.

(i) Variables can take more than one value or belong to more than one category. For example, a school can be described by the salaries of its three most highly paid teachers or by the foreign languages taught in it.

(ii) Variables can be defined to belong to a specified interval of values. For example, the time in minutes taken by an individual to travel to work may be known only to lie in the interval [20, 30], and the heights in centimetres of the individual's children may belong to the interval [150, 180].

(iii) Variables can be specified by a (continuous or discrete) distribution. For example, the height of 18-20 year-old males in Britain could be represented by a normal distribution with a specified mean and variance; the number of service calls made by a tradesman in a day could be: 3 with probability 0.5, 4 with probability 0.4, and 5 with probability 0.1. Alternatively, probability distributions could be replaced by fuzzy measures or 'possibilist' assertions (Diday, 1995).

The fact that some of these variables can be represented symbolically (e. g. as a bar chart or histogram) has led to their being referred to as 'symbolic variables'. As illustrated by the examples, such variables can be used in the description of (a) individual objects and (b) classes of individual objects. Thus, in (ii), the fact that the travelling time of the individual lies between 20 and 30 minutes is taken as indicating uncertainty about the precise length of time taken. However, the variable describing the children's heights has been derived from the information provided by the set of individual children. Similarly, in (iii), the tradesman is an individual object whose number of service calls cannot be exactly specified, whereas the information about the height of British 18-20 year-old males is derived from a set of such individuals. Individual objects (resp., classes of individual objects) described by symbolic variables are referred to as first order (resp., second order) symbolic objects.

As discussed earlier, a first stage in the classification of a set of objects is often the construction of a set of (dis)similarities between each pair of objects. Two measures of dissimilarity relevant for symbolic objects which are described by variables of type $(i)$ and $(ii)$ are presented below; measures of (dis)similarity relevant for symbolic objects described by variables of type $(iii)$ are given in Section 2.2.5.

Suppose that the description of the $k$th symbolic variable, $V_k$, for the $i$th object $(i = 1, ..., n; k = 1, ..., p)$ is given by $V_k = v_{ik}$, where $v_{ik}$ could be an interval $[v_{ikl}, v_{iku}]$ for a quantitative variable or a set of (possibly ordered) states $\{v_{ikl}, ..., v_{iku}\}$ of a categorical variable. It is convenient to define the Cartesian join ($\oplus$) and meet ($\otimes$) of the values $v_{ik}$ and $v_{jk}$. The Cartesian join, $v_{ik} \oplus v_{jk}$, is defined by:

$$
v_{ik} \oplus v_{jk} \equiv \begin{cases} [\min\{v_{ikl}, v_{jkl}\}, \max\{v_{iku}, v_{jku}\}] \text{ if } V_k \text{ is a} \\ \qquad\qquad \text{quantitative or ordinal variable;} \\ v_{ik} \cup v_{jk} \text{ if } V_k \text{ is a nominal variable.} \end{cases}
$$

$$(5.22)$$

For all these types of variable, the Cartesian meet, $v_{ik} \otimes v_{jk}$, is defined by:

$$
v_{ik} \otimes v_{jk} \equiv v_{ik} \cap v_{jk}. \tag{5.23}
$$

Gowda and Diday (1991) define the contribution $d_{ijk}$ of $V_k$ to the dissimilarity between the $i$th and $j$th symbolic objects to be the sum of differences in position $(d_{ijk}^{(P)})$, span $(d_{ijk}^{(S)})$, and content $(d_{ijk}^{(C)})$:

$$
d_{ijk} = d_{ijk}^{(P)} + d_{ijk}^{(S)} + d_{ijk}^{(C)}. \tag{5.24}
$$

Let $\mu(v_{ik})$ denote the length of the interval $v_{ik}$ if $V_k$ is a quantitative variable, or the number of possible values included in the set $v_{ik}$ if $V_k$ is a categorical variable. Define $\mu(V_k)$ $(< \infty)$ to be the length of the maximum interval for (or maximum number of categories in) $V_k$. A slightly modified version of Gowda and Diday's (1991) definitions is as follows:

$$d_{ijk}^{(P)} \equiv \begin{cases} |v_{ikl} - v_{jkl}|/\mu(V_k) \text{ if } V_k \text{ is a quantitative variable} \\ 0 \text{ if } V_k \text{ is a categorical variable} \end{cases}$$

$$\text{(5.25)}$$

$$d_{ijk}^{(S)} \equiv |\mu(v_{ik}) - \mu(v_{jk})|/\mu(v_{ik} \oplus v_{jk}) \qquad \text{(5.26)}$$

$$d_{ijk}^{(C)} \equiv [\mu(v_{ik}) + \mu(v_{jk}) - 2\mu(v_{ik} \otimes v_{jk})]/\mu(v_{ik} \oplus v_{jk}). \qquad \text{(5.27)}$$

The contribution is defined to be zero if both the numerator and denominator in any of the definitions (5.25) − (5.27) are zero, for example if $V_k$ can take only one value or both 'intervals' are the same single value.

An alternative measure of pairwise dissimilarity proposed by Ichino and Yaguchi (1994) is defined by:

$$d_{ijk} \equiv [\mu(v_{ik} \oplus v_{jk}) - \mu(v_{ik} \otimes v_{jk}) + \gamma\nu(v_{ik}, v_{jk})]/\mu(V_k), \quad \text{(5.28)}$$

where

$$\nu(v_{ik}, v_{jk}) \equiv 2\mu(v_{ik} \otimes v_{jk}) - \mu(v_{ik}) - \mu(v_{jk}). \qquad \text{(5.29)}$$

The value of the parameter $\gamma(0 \leq \gamma \leq \frac{1}{2})$ is important when $\mu(v_{ik} \otimes v_{jk}) = 0$. Ichino and Yaguchi (1994) suggest that one of the simplest choices is to use $\gamma = \frac{1}{2}$, when

$$d_{ijk} \equiv [\mu(v_{ik} \oplus v_{jk}) - (\mu(v_{ik}) + \mu(v_{jk}))/2]/\mu(V_k). \qquad \text{(5.30)}$$

Variables can also be structured or related to one another:

- Variables may have a hierarchical structure to their categories or values. For example, a colour could be described as 'dark blue' or 'black', or the word 'dark' could be used to refer to either of these colours.

- There may be dependencies between variables. One example is provided by the conditionally-present variables discussed in Section 2.3.3.

Ichino and Yaguchi (1994) and de Carvalho (1998) discuss the construction of measures of dissimilarity between symbolic objects described by these types of variable.

The overall measure of the dissimilarity between the $i$th and $j$th symbolic objects can be defined to be

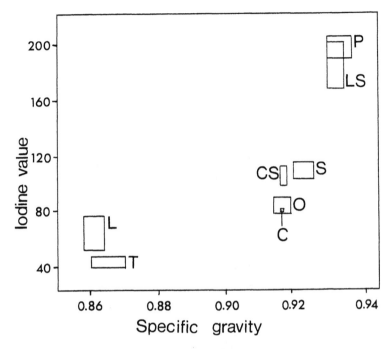

Figure 5.9 *A plot of the fats and oils symbolic objects on their first and third variables. Abbreviations for the objects are explained in Table 5.6*

$$d_{ij} \equiv [\Sigma_{k=1}^{p}(w_k d_{ijk})^{\lambda}]^{1/\lambda} \ (\lambda \geq 1), \qquad (5.31)$$

where $w_k \geq 0 \ (k = 1, ..., p)$ are non-negative weights associated with the variables.

As an illustration, consider the information about eight fats and oils contained in Table 5.6, which is extracted from Gowda and Diday (1991) and Ichino and Yaguchi (1994). Each of these symbolic objects is described by four quantitative variables which take values within an interval, and one nominal variable which takes a finite set of values. Fig. 5.9 shows a plot of the data on the first and third of these variables.

Table 5.7 summarizes information necessary for evaluating the dissimilarity measure (5.30) between the first two objects, linseed oil and perilla oil; $\mu(V_k)$ is defined to be the range of values taken by (or number of categories of) the $k$th variable in the complete

Table 5.6 *Information about eight fats and oils.*

| Name | Code | Specific gravity (g/cm$^3$) | Freezing point (deg C) | Iodine value | Saponification value | Major fatty acids* |
|---|---|---|---|---|---|---|
| Linseed oil | LS | [0.930, 0.935] | [−27, −8] | [170, 204] | [118, 196] | L, Ln, M, O, P |
| Perilla oil | P | [0.930, 0.937] | [−5, −4] | [192, 208] | [188, 197] | L, Ln, O, P, S |
| Cotton seed | CS | [0.916, 0.918] | [−6, −1] | [99, 113] | [189, 198] | L, M, O, P, S |
| Sesame oil | S | [0.920, 0.926] | [−6, −4] | [104, 116] | [187, 193] | A, L, O, P, S |
| Camelia | C | [0.916, 0.917] | [−21, −15] | [80,82] | [189, 193] | L, O |
| Olive oil | O | [0.914, 0.919] | [0, 6] | [79, 90] | [187, 196] | L, O, P, S |
| Beef tallow | T | [0.860, 0.870] | [30, 38] | [40,48] | [190, 199] | C, M, O, P, S |
| Lard | L | [0.858, 0.864] | [22, 32] | [53, 77] | [190, 202] | L, Lu, M, O, P, S |

* Codes for fatty acids: A = arachic acid; C = capric acid; L = linoleic acid; Ln = Linolenic acid; Lu = lauric acid; M = myristic acid; O = oleic acid; P = palmitic acid; S = stearic acid.

Table 5.7 *Information allowing evaluation of a measure of the dissimilarity between the first and second symbolic objects described in Table 5.6.*

| $k$ | $v_{1k} \oplus v_{2k}$ | $\mu(v_{1k} \oplus v_{2k})$ | $[\mu(v_{1k}) + \mu(v_{2k})]/2$ | $\mu(V_k)$ |
|---|---|---|---|---|
| 1 | [0.930, 0.937] | 0.007 | (0.005 + 0.007)/2 | 0.079 |
| 2 | [−27, −4] | 23 | (19 + 1)/2 | 65 |
| 3 | [170, 208] | 38 | (34 + 16)/2 | 168 |
| 4 | [118, 197] | 79 | (78 + 9)/2 | 84 |
| 5 | {L,Ln,M,O,P,S} | 6 | (5 + 5)/2 | 9 |

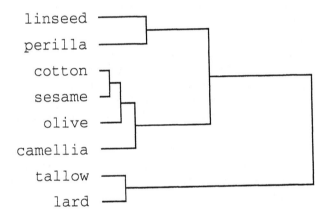

Figure 5.10 *A complete link analysis of the fats and oils data.*

data set. If the weights $w_k$ $(k = 1, ..., p)$ and the exponent $\lambda$ are all defined to be 1,

$$d_{12} = 0.001/0.079 + 13/65 + 13/168 + 35.5/84 + 1/9 = 0.824.$$

Given a relevant measure of pairwise dissimilarity between symbolic objects, classifications of them can be obtained using standard algorithms that analyse dissimilarity matrices, as described in Chapters 3 and 4. For example, the results of a complete link analysis of the fats and oils data set based on the dissimilarity measure (5.30) are shown in Fig. 5.10.

In such an approach, once the matrix of dissimilarities has been constructed, the classification is obtained without any further attention being paid to the fact that it is symbolic data rather than more conventional data that are being investigated. A more elaborate investigation, which has more in common with conceptual clustering, involves specifying new symbolic objects whenever there is a change in class membership, e.g. in the course of an analysis using an agglomerative algorithm. Thus, the second order symbolic object corresponding to the class comprising the $i$th and $j$th symbolic objects can be defined to have $k$th symbolic variable $v_{ik} \oplus v_{jk}$ $(k = 1, ..., p)$; the second column of Table 5.7 presents this information for the symbolic object {linseed oil, perilla oil}. The dissimilarities between this new symbolic object and all other remaining symbolic objects can then be recalculated (Gowda and Diday, 1994).

It is important to provide informative descriptions for classes of objects, although these might be specified only at the end of an investigation, after all the classes of objects have been identified; symbolic objects provide one approach to obtaining such descriptions. A further discussion of this topic is postponed to Section 7.3.

A fuller account of the analysis of symbolic data is presented by Bock et al. (1999), and a recent project to develop software for classification and other analyses of symbolic data is described by Hébrail (1998).

## 5.6 Partitions of partitions

In another non-standard classification problem, the basic data consist of partitions of a set of objects; an example is provided by the kinship terms data. In order to summarize these data, it can be informative to obtain a *consensus* partition, which summarizes the information provided by the 85 subjects. However, it is possible that different subjects have markedly different views about the perceived similarities of kinship terms, and it can be relevant to partition the subjects into homogeneous classes. Some terminology intended to avoid confusion is introduced: a partition of a set of objects (e.g. kinship terms) is called a *primary* partition; a partition of a set of primary partitions is called a *secondary* partition. The aim is to obtain a secondary partition of a set of primary partitions into disjoint classes such that primary partitions in the same class are perceived as similar to one another; associated with each of the classes is a consensus partition which summarizes the primary partitions in the class (Gordon and Vichi, 1998).

Let there be $m$ primary partitions, $\{P_1, ..., P_m\}$, of the same set of $n$ objects. The set of primary partitions can be specified by a three-way binary array $\mathbf{C} \equiv (c_{ijk})$, where $c_{ijk} = 1$ if the $i$th and $j$th objects belong to the same class in $P_k$, and 0 if they do not $(i, j = 1, ..., n; k = 1, ..., m)$. A common measure of the difference between two primary partitions, $P_k$ and $P_l$, is the number of pairs of objects belonging to different classes in the two partitions:

$$\Delta(P_k, P_l) \equiv \Sigma_{1 \le i < j \le n} (c_{ijk} - c_{ijl})^2. \qquad (5.32)$$

The *median* consensus partition of the set of $m$ primary partitions is defined to be a partition $M$ which minimizes $\Sigma_{k=1}^{m} \Delta(P_k, M)$ (Régnier, 1965; Leclerc and Cucumel, 1987). If $M$ is specified by the binary matrix $\mathbf{Y} \equiv (y_{ij})$, where $y_{ij} = 1$ (resp., 0) if the $i$th and $j$th objects belong (resp., do not belong) to the same class in the partition $M$, the median partition problem can be formulated in terms of finding a partition specified by $\mathbf{Y}$ which minimizes

$$F(\mathbf{Y}) \equiv \Sigma_{k=1}^{m} \Sigma_{1 \le i < j \le n} (c_{ijk} - y_{ij})^2. \qquad (5.33)$$

Since the elements of $\mathbf{C}$ and $\mathbf{Y}$ are all either 0 or 1, it follows (Régnier, 1965; Marcotorchino and Michaud, 1982) that

$$
\begin{aligned}
F(\mathbf{Y}) &= \Sigma_{k=1}^{m} \Sigma_{1 \le i < j \le n} c_{ijk} - \Sigma_{1 \le i < j \le n} y_{ij} \Sigma_{k=1}^{m} (2c_{ijk} - 1) \\
&= \text{constant} - \Sigma_{1 \le i < j \le n} b_{ij} y_{ij},
\end{aligned}
$$

where

$$b_{ij} = 2\Sigma_{k=1}^{m} c_{ijk} - m. \qquad (5.34)$$

Some side conditions are necessary to ensure that $\mathbf{Y}$ specifies a partition. Thus, the aim is to maximize

$$\Sigma_{i=1}^{n-1} \Sigma_{j=i+1}^{n} b_{ij} y_{ij} \qquad (5.35)$$

subject to

$$y_{ij} \in \{0, 1\} \quad (1 \le i < j \le n) \qquad (5.36)$$

$$\left.\begin{array}{l} y_{ik} + y_{jk} - y_{ij} \leq 1 \\ y_{ij} + y_{jk} - y_{ik} \leq 1 \\ y_{ik} + y_{ij} - y_{jk} \leq 1 \end{array}\right\} \quad (1 \leq i < j < k \leq n). \tag{5.37}$$

This integer linear programming problem can be relaxed to a linear programming problem by replacing condition (5.36) by

$$0 \leq y_{ij} \leq 1 \ (1 \leq i < j \leq n). \tag{5.38}$$

A solution to the linear programming problem which has all $y_{ij} = 0$ or 1 is also a solution to the integer linear programming problem. It has been found in practice that the linear programming problem often, but not always, has a 0/1 solution (Grötschel and Wakabayashi, 1989). If the solution is not integral, there is the need for more elaborate algorithms, such as those reviewed by Hansen, Jaumard and Sanlaville (1994); such algorithms were not required to obtain the results described in this section.

When this methodology was applied to the kinship terms data, the median consensus partition was found to be:

{grandfather, grandmother}, {grandson, granddaughter}, {brother, sister}, {father, mother, son, daughter}, {nephew, niece}, {uncle, aunt, cousin}.

In order to obtain a secondary partition of the set of primary partitions into $c$ classes, it is necessary to define a set of matrices $\mathbf{Y_r} \equiv (y_{ijr})$, where $y_{ijr} = 1$ (resp., 0) if the $i$th and $j$th objects belong (resp., do not belong) to the same class in the consensus partition associated with the $r$th class, $C_r$ $(i, j = 1, ..., n; r = 1, ..., c)$. The aim is then to minimize

$$\Sigma_{r=1}^{c} \Sigma_{k \in C_r} \Sigma_{1 \leq i < j \leq n} \left(c_{ijk} - y_{ijr}\right)^2. \tag{5.39}$$

An optimal solution can be sought using the following two-stage iterative relocation algorithm:

1. A median partition is identified for each class in the current secondary partition.

2. Each primary partition is assigned to a class to whose consensus partition it is closest.

These steps are repeated until there is no further change in the secondary partition. This algorithm has features in common with the $k$-means algorithm described in Section 3.2, and − like it − cannot be guaranteed to find globally optimal solutions. Further,

Table 5.8 *Consensus partitions of the kinship terms data when the 85 subjects are partitioned into three classes.*

| Class | Consensus partition |
|-------|---------------------|
| $C_1$ | {grandfather, grandmother, grandson, granddaughter}, {brother, sister}, {father, mother, son, daughter}, {nephew, niece, uncle, aunt, cousin} |
| $C_2$ | {grandfather, grandmother}, {grandson, granddaughter}, {brother, sister}, {father, mother}, {son, daughter}, {nephew, niece}, {uncle, aunt}, {cousin} |
| $C_3$ | {grandfather, grandson, brother, father, son, nephew, uncle}, {grandmother, granddaughter, sister, mother, daughter, niece, aunt}, {cousin} |

Table 5.9 *Assignment of the 85 subjects into the three classes whose consensus partitions are given in Table 5.8.*

| Class | Subject numbers |
|-------|-----------------|
| $C_1$ | 9, 10, 12, 14–16, 19–32, 59, 62–71, 73–79 |
| $C_2$ | 1–8, 13, 17, 18, 33–45, 47–57, 60, 61 |
| $C_1$ or $C_2$ | 11, 46, 58, 72 |
| $C_3$ | 80–85 |

not only the class membership, but also the consensus partitions, need not be uniquely defined.

The results obtained when the kinship terms data were partitioned into three classes, summarized in Tables 5.8 and 5.9, are in good agreement with the earlier analysis of these data reported in Section 4.5. The main feature of interest is that while subjects placed in the first two classes have grouped together equivalent terms referring to individuals of different gender (e.g. {father, mother}, {brother, sister}), the six subjects in the third class have categorized the kinship terms solely by gender (the word 'cousin' having an ambiguous gender in the English language). Further, there are four subjects whose partitions are equidistant from the consensus partitions associated with the first two classes.

In the theory presented thus far in this section, no restrictions have been placed on what constitutes an allowable secondary partition. However, such constraints can be imposed using methodology described in Section 5.2; an application to obtaining secondary partitions of a set of time-ordered primary partitions is described by Gordon and Vichi (1998).

# CHAPTER 6

# Graphical representations

## 6.1 Introduction

This chapter describes several methods of representing a set of objects by a set of points in a low-dimensional space, one point for each object, so that objects that are similar to one another are represented by points that are close together. Two-dimensional representations have been most commonly obtained, although three-dimensional physical constructs and diagrams drawn in perspective have also been used. Such configurations of points can then be examined to see if they fall into compact and isolated classes of points.

This approach has the advantage that class structure is not imposed on the set of objects, as is the case with methodology described in earlier chapters. However, the assessment by eye of configurations of points can be highly subjective, with different observers reaching different conclusions. Further, it might not be possible to capture in a low-dimensional representation all of the variability present in a data set. It is thus useful to supplement such graphical representations by incorporating into them the results of other analyses; for example, several Figures in Chapter 3 provide information about a partition, and Fig. 3.8 superimposes a minimum spanning tree on a graphical representation. Such combined analyses reduce the element of subjectivity in assessing results, while allowing the possibility of checking if inappropriate structure is being imposed on the data. If similar conclusions would be reached from the separate analyses, one can have more confidence in the accuracy of the results.

Graphical representations have been widely used in many disciplines, and different terminology has been employed to refer to similar methodology. For example, the material in the next two sections of this chapter has been referred to by the terms 'multidimensional scaling' in psychology and 'ordination' in ecology.

This chapter describes several ways of obtaining a graphical rep-

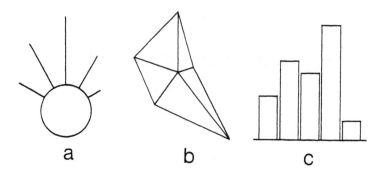

Figure 6.1 *Several icons representing values of five variables: (a) met-roglyph, (b) star or polygon, (c) profile.*

resentation of a set of objects by optimizing a goodness-of-fit cri-terion. The values taken by these criteria provide measures of the extent to which the low-dimensional configuration of points fails to provide an adequate representation of the data. Various ways have been proposed of including information about higher dimensions in two-dimensional configurations. For example, coordinate values on a third dimension can be represented by the colour or size of points, or by their brightness in a visual display unit. Points can also be replaced by icons located at the same positions. Fig. 6.1 portrays several different icons which have been used to describe an object in terms of a set of variables but which can also be used to repre-sent higher coordinate values. The size of a variable is shown by the length of the corresponding ray in the metroglyph (Anderson, 1960) or star (Kleiner and Hartigan, 1981) or by the height of a box in the profile (Bertin, 1983); other icons are discussed by Kleiner and Hartigan (1981) and Tukey and Tukey (1981b). The ability to assimilate rapidly the information contained in such icons requires practice and can be hindered if they contain too much detail.

An outline of the contents of the remainder of the chapter is as follows. The next two sections describe principal coordinates analysis and non-metric multidimensional scaling. These are both methods of obtaining a configuration of points whose interpoint distances approximate a given dissimilarity matrix. The methods differ in several respects, notably in the assumptions made about how the dissimilarities are related to the distances and in the cri-terion to be optimized. Section 6.4 describes a miscellaneous collection of other methodology that has been proposed for obtaining and assessing configurations of points. The final section describes methodology for obtaining simultaneous graphical representations both of a set of objects and of the set of variables describing the objects; thus, points or vectors represent objects and variables, and the joint graphical representation allows one to assess not only similarities between objects and similarities between variables, but also interactions between the set of objects and the set of variables.

## 6.2 Principal coordinates analysis

Given a set of $n$ points in $p$-dimensional Euclidean space, $\{P_i\,(i = 1,...,n)\}$ with coordinates $\{x_{ik}\,(i = 1,...,n; k = 1,...,p)\}$, it is straightforward to evaluate the squared Euclidean distance $\Delta_{ij}$ between any pair of points, $P_i$ and $P_j$, by use of the formula

$$\Delta_{ij} \;=\; \Sigma_{k=1}^{p}(x_{ik} - x_{jk})^2. \tag{6.1}$$

This section addresses the converse problem: given an $n \times n$ matrix of values $(\Delta_{ij})$ which are known (or assumed) to represent the squared distances within a set of $n$ points in some Euclidean space, obtain the coordinates of these points. The following procedure was developed by Schoenberg (1935), Young and Householder (1938) and Torgerson (1952), and is sometimes referred to as 'classical scaling'. It was popularized, under the title 'principal coordinates analysis', by Gower (1966), who noted links between it and other statistical methodology.

A configuration of points provides the same set of interpoint distances if its axes are translated and rotated/reflected. The first indeterminacy is removed by requiring the centroid of the set of points to lie at the origin of coordinates, i.e.

$$\Sigma_{i=1}^{n} x_{ik} \;=\; 0\,(k = 1, 2, ...). \tag{6.2}$$

The axes of the solution will be shown to be principal component axes.

The values $\{x_{ik}\}$ are obtained from $(\Delta_{ij})$ by a two-stage procedure. The first stage involves transforming the matrix $(\Delta_{ij})$ into an inner-product matrix $\mathbf{B} \equiv (b_{ij})$, where

$$b_{ij} \;\equiv\; \Sigma_k x_{ik} x_{jk}\,(i, j = 1, ..., n) \tag{6.3}$$

and, from Equation (6.2),

$$\Sigma_{j=1}^{n} b_{ij} = 0 = \Sigma_{i=1}^{n} b_{ij}.$$

It can be shown that

$$b_{ij} = -\tfrac{1}{2}(\Delta_{ij} - \Delta_{i.} - \Delta_{.j} + \Delta_{..}), \qquad (6.4)$$

where

$$\Delta_{i.} = \Delta_{.i} \equiv \Sigma_{j=1}^{n}\Delta_{ij}/n \ (i = 1, ..., n)$$
$$\Delta_{..} \equiv \Sigma_{i=1}^{n}\Sigma_{j=1}^{n}\Delta_{ij}/n^{2}.$$

Equation (6.4) can be proved by noting from Equations (6.1) and (6.3) that

$$\Delta_{ij} = b_{ii} - 2b_{ij} + b_{jj} \qquad (6.5)$$

and by substituting for $\Delta_{ij}$ etc. in the right-hand side of Equation (6.4).

Since $\mathbf{B}$ is a real symmetric matrix, it can be diagonalized:

$$\mathbf{B} = \mathbf{V}\boldsymbol{\Lambda}\mathbf{V}', \qquad (6.6)$$

where the columns of the orthogonal matrix $\mathbf{V} \equiv (\mathbf{v_1}, \mathbf{v_2}, ..., \mathbf{v_n})$ contain the eigenvectors corresponding to the eigenvalues ($\lambda_1$, $\lambda_2$, ..., $\lambda_n$) appearing in the $(n \times n)$ diagonal matrix $\boldsymbol{\Lambda}$. The eigenvectors have unit length ($\mathbf{v_i'}\mathbf{v_i} = 1 \ (i = 1, ..., n)$), and the eigenvalues are assumed to be numbered in non-increasing order of magnitude:

$$\lambda_1 \geq \lambda_2 \geq ... \geq \lambda_n.$$

If $\mathbf{X} \equiv (x_{ik})$, Equation (6.3) can equivalently be expressed as

$$\mathbf{B} = \mathbf{X}\mathbf{X}', \qquad (6.7)$$

and a comparison of Equations (6.6) and (6.7) shows that the unknown matrix $\mathbf{X}$ can be obtained from

$$\mathbf{X} = \mathbf{V}\boldsymbol{\Lambda}^{\frac{1}{2}}$$
$$= (\sqrt{\lambda_1}\mathbf{v_1}, \sqrt{\lambda_2}\mathbf{v_2}, ..., \sqrt{\lambda_n}\mathbf{v_n}). \qquad (6.8)$$

If the set of interpoint distances can be recovered from an $s$-dimensional configuration, where $s < n$, $\lambda_j = 0 \ (j = s + 1, ..., n)$.

The solution is indeterminate to the extent that arbitrary reflections in the origin are allowed.

As an illustration, the following matrix contains the squared distances within a set of five points, $\{P_i \ (i = 1, ..., 5)\}$:

$$(\Delta_{ij}) = \begin{pmatrix} 0 & 4 & 5 & 16 & 20 \\ 4 & 0 & 5 & 20 & 16 \\ 5 & 5 & 0 & 5 & 5 \\ 16 & 20 & 5 & 0 & 4 \\ 20 & 16 & 5 & 4 & 0 \end{pmatrix}.$$

Using Equation (6.4), the corresponding inner product matrix is evaluated as

$$\mathbf{B} = \begin{pmatrix} 5 & 3 & 0 & -3 & -5 \\ 3 & 5 & 0 & -5 & -3 \\ 0 & 0 & 0 & 0 & 0 \\ -3 & -5 & 0 & 5 & 3 \\ -5 & -3 & 0 & 3 & 5 \end{pmatrix}.$$

The matrix $\mathbf{B}$ has two positive eigenvalues and three zero eigenvalues. The positive eigenvalues, together with their standardized eigenvectors, are:

$$\lambda_1 = 16, \ \mathbf{v_1} = (-\tfrac{1}{2}, -\tfrac{1}{2}, 0, \tfrac{1}{2}, \tfrac{1}{2})';$$
$$\lambda_2 = 4, \ \mathbf{v_2} = (\tfrac{1}{2}, -\tfrac{1}{2}, 0, \tfrac{1}{2}, -\tfrac{1}{2})'.$$

Hence,

$$\mathbf{X} = (\sqrt{\lambda_1}\mathbf{v_1}, \sqrt{\lambda_2}\mathbf{v_2}) = \begin{pmatrix} -2 & 1 \\ -2 & -1 \\ 0 & 0 \\ 2 & 1 \\ 2 & -1 \end{pmatrix},$$

where the positive square roots of $\lambda_1$ and $\lambda_2$ have been taken. This configuration is sketched in Fig. 6.2; it is seen that the centroid of the points is at the origin and that the squared distances between the points are as given by the matrix $(\Delta_{ij})$.

There are links between principal coordinates analysis and principal components analysis (Gower, 1966), as described below. There are two main formulations of principal components analysis (Jolliffe, 1986). The first (Pearson, 1901) aims to project a configuration of points in $p$ dimensions onto an $s$-dimensional hyperplane

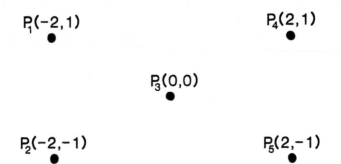

Figure 6.2 *Graphical representation of the set of points* $\{P_i \ (i = 1, ..., 5)\}$ *which have squared interpoint distances given by the matrix* $(\Delta_{ij})$.

$(s < p)$ so that the sum of the squares of the projections is minimized. This involves a rigid rotation of the coordinate axes to new axes (which turn out to be the principal components) such that for all $s < p$, the first $s$ axes define the best (in the sense of least squares) $s$-dimensional hyperplane. The second formulation (Hotelling, 1933) seeks linear combinations of the original variables, called principal components, which have maximum variance subject to being uncorrelated with one another.

Let the coordinates of the points be held in the $(n \times p)$ matrix $\mathbf{X}$. By the singular value decomposition theorem (Eckart and Young, 1936; Johnson, 1963),

$$\mathbf{X} = \mathbf{T\Theta U'} = \Sigma_{j=1}^{r} \theta_{jj} \mathbf{t_j} \mathbf{u_j'}, \qquad (6.9)$$

where $r(\leq min(n, p))$ denotes the rank of $\mathbf{X}$; $\mathbf{T} \equiv (\mathbf{t_1}, ..., \mathbf{t_n})$ and $\mathbf{U} \equiv (\mathbf{u_1}, ..., \mathbf{u_p})$ are orthogonal matrices; and $\mathbf{\Theta}$ is an $(n \times p)$ diagonal matrix whose only non-zero elements are $\{\theta_{jj} \ (j = 1, ..., r)\}$, with

$$\theta_{11} \geq \theta_{22} \geq ... \geq \theta_{rr} > 0.$$

In the following, it is assumed that $\mathbf{X}$ is centred at the origin, implying that $r \leq min(n - 1, p)$.

For $s < r$, the least squares $s$-dimensional approximation of $\mathbf{X}$ is given by

$$\mathbf{X_s} \equiv \Sigma_{j=1}^{s} \theta_{jj} \mathbf{t_j} \mathbf{u_j'},$$

with sum of squared projections equal to

$$\text{trace}[(\mathbf{X} - \mathbf{X_s})'(\mathbf{X} - \mathbf{X_s})]$$
$$= \text{trace}\{(\Sigma_{i=s+1}^{r}\theta_{ii}\mathbf{u_i}\mathbf{t_i'})(\Sigma_{j=s+1}^{r}\theta_{jj}\mathbf{t_j}\mathbf{u_j'})\}$$
$$= \Sigma_{j=s+1}^{r}\theta_{jj}^2.$$

The second formulation of principal components analysis is based on an eigenanalysis of the sample covariance matrix, $\mathbf{S} \equiv \mathbf{W}/(n-1)$, where

$$\mathbf{W} \equiv \mathbf{X'X} = \mathbf{U\Theta'T'T\Theta U'} = \mathbf{U\tilde{\Lambda}U'},$$

and $\tilde{\Lambda}$ is a $(p \times p)$ diagonal matrix with $j$th diagonal element

$$\tilde{\lambda}_j = \left\{ \begin{array}{ll} \theta_{jj}^2 & (1 \leq j \leq r) \\ 0 & (r+1 \leq j \leq p; \text{ if } p > r). \end{array} \right.$$

The principal components are defined by the eigenvectors of $\mathbf{W}$, stored in the columns of the matrix $\mathbf{U}$. Suppose now that $\mathbf{X}$ is defined by Equation (6.8), this implying that $r < n$. Comparing Equations (6.8) and (6.9), it is seen that if $\mathbf{T}$ is identified with $\mathbf{V}$ and the set $\{\lambda_j \ (j = 1, ..., n)\}$ is defined by

$$\lambda_j = \left\{ \begin{array}{ll} \tilde{\lambda}_j & (1 \leq j \leq r) \\ 0 & (r+1 \leq j \leq n), \end{array} \right.$$

then $\mathbf{U}$ is an identity matrix, i.e. the coordinate axes of $\mathbf{X}$ are the principal components. Hence, for any value of $s < n$, an optimal least squares configuration of points in $s$ dimensions is obtained from the first $s$ columns of the matrix $\mathbf{X}$ defined in Equation (6.8).

For this result to hold, the matrix $(\Delta_{ij})$ must exactly reproduce squared interpoint distances in some Euclidean space, implying that all the eigenvalues of the matrix $\mathbf{B}$ are non-negative. However, the approach can be used more widely: if $(\Delta_{ij})$ are approximately squared distances within some configuration of points, one might expect any negative eigenvalues to be small in magnitude, and such expectations receive support from perturbation studies (Mardia, 1978; Sibson, 1979). Hence, one can use Equations (6.4) and (6.6) to carry out an eigenanalysis of the 'inner product' matrix $\mathbf{B}$ and obtain an approximation to the configuration using dimensions corresponding to positive eigenvalues (or only the first few of them).

This methodology can be applied to any dissimilarity matrix, whether provided directly or constructed using one of the measures described in Chapter 2. It is usually preferable for dissimilarities between pairs of objects to be represented by distances, rather than squared distances, between pairs of points in a configuration; this can be achieved by squaring the elements of the dissimilarity matrix before carrying out the analysis.

By analogy with principal components analysis, the adequacy of an $s$-dimensional configuration can be measured by

$$A(s) \equiv \Sigma_{i=1}^s \lambda_i / \Sigma_{j=1}^n |\lambda_j|. \tag{6.10}$$

However, if any of the negative eigenvalues is large in magnitude, any such configuration should be regarded with suspicion.

If dissimilarities $(d_{ij})$ represent exactly interpoint distances within a set of points in $p$-dimensional space and $(\hat{d}_{ij})$ denote distances within the same set of points in $s$-dimensional space $(s < p)$, the form of Equation (6.8) shows that principal coordinates/components analysis provides a projection of the original configuration of points into $s$ dimensions which minimizes

$$\Sigma_{1 \le j < i \le n}(d_{ij}^2 - \hat{d}_{ij}^2).$$

Because the $(\hat{d}_{ij})$ are obtained from a projected configuration, $\hat{d}_{ij} \le d_{ij}\,(i, j = 1, ..., n)$. Several authors have suggested the direct minimization of other measures of the divergence between the sets $(d_{ij})$ and $(\hat{d}_{ij})$, for example (Sammon, 1969)

$$\Sigma_{1 \le j < i \le n} w_{ij}(d_{ij} - \hat{d}_{ij})^2, \tag{6.11}$$

where $(w_{ij})$ are weighting factors which can be selected so as to bias the configuration towards more accurate representation of some of the distances, e.g. those that are larger. An iterative approximating algorithm is used to estimate the $(\hat{d}_{ij})$ which minimize Expression (6.11)

Several of the Figures in Chapter 3 show graphical representations of the Abernethy Forest data sets, providing two-dimensional plots of nine-dimensional data with some supplementary information superimposed: thus, Figs. 3.2 and 3.8 portray the 1974 data set, and Fig. 3.3 analyses both data sets together. The values of $A(2)$ (Expression (6.10)) for the two configurations are 0.79 and 0.78, respectively, indicating that in each case the great bulk of the variability in the nine-dimensional data set has been captured in two dimensions. Given that the original pattern matrices were

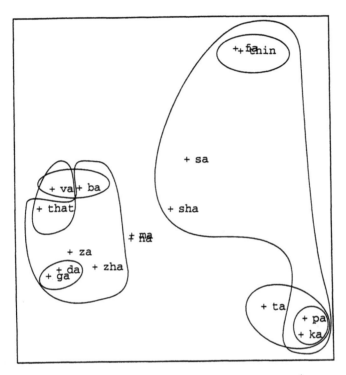

Figure 6.3 *A principal coordinates analysis of the acoustic confusion data; the overlapping classes specified in Table 5.2 are shown by the bounding curves.*

available, it is more efficient to obtain these plots by a principal components analysis of the (9 × 9) sample covariance matrices, rather than by principal coordinates analyses of the (49 × 49) and (90 × 90) inner product matrices. However, when the information about the relationships within a set of objects is contained in a (dis)similarity matrix rather than a pattern matrix, this choice is not available.

The acoustic confusion data are described by a similarity matrix, and principal coordinates analysis is used to provide a graphical representation of them. The matrix of 'squared distances' $(\Delta_{ij})$ is obtained from the matrix of similarities $(s_{ij})$ by the transformation

$$\Delta_{ij} = c - s_{ij} \; (i, j = 1, ..., n).$$

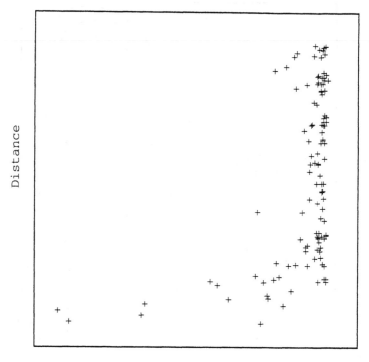

Dissimilarity

Figure 6.4 *A plot of the interpoint distances in Fig. 6.3 against* $(\Delta_{ij}^{1/2})$ *for the acoustic confusion data.*

After some investigation, the constant $c$ was chosen to be $\frac{1}{2}$, slightly larger than the largest element of the similarity matrix. A more natural choice for $c$ might seem to be $c = 1$, but this provided a configuration with a very low value of $A(2)$. The best two-dimensional representation of the data, which still accounts for only 29% of the variability, is shown in Fig. 6.3. The boundaries of the overlapping classes found by Arabie and Carroll (1980) and defined in Table 5.2 are also plotted in the Figure, indicating a reasonable correspondence between the two sets of results. However, a plot of the interpoint distances against $(\Delta_{ij}^{\frac{1}{2}})$, given in Fig. 6.4, is far from linear, suggesting that a better representation can be obtained by assuming a different relationship between dissimilarities and distances.

## 6.3 Non-metric multidimensional scaling

In principal coordinates analysis, it is assumed that the (squared) dissimilarity between each pair of objects is approximately the squared distance between the corresponding pair of points in a graphical representation. This relationship holds exactly in the small worked example given in the previous section, but in general one might expect dissimilarities and distances to be much less simply related, as is the case for the acoustic confusion data. Methodology has been proposed for obtaining graphical representations in which the relationship between dissimilarity and distance is modelled by a more general parametric function (Shepard, 1974; Ramsay, 1977; Critchley, 1978) or by monotone spline functions (Ramsay, 1982). This section describes the method of non-metric multidimensional scaling developed by Shepard (1962a, b) and Kruskal (1964a, b), in which it is assumed only that dissimilarities and distances are monotonically related.

In this methodology, the only information about the dissimilarities that is used is their rank ordering; it is not assumed that they contain any metric information, hence the qualifying term 'non-metric'. However, the result of the analysis is a configuration of points, which does contain metric information, and it might be more appropriate to refer to the methodology as 'ordinal scaling'.

Shepard (1962a, b) presented a heuristic algorithm for seeking a configuration whose interpoint distances are approximately monotonically related to a given set of dissimilarities; his approach did not involve an explicit minimization of a function measuring the departure from perfect monotonicity between dissimilarities and distances. These ideas were formalized by Kruskal (1964a, b), using the method of least squares monotone regression. This is a method of comparing two sequences of real numbers, and it is described in this general context before its application to comparing sets of dissimilarities and distances is illustrated.

Assume that $\mathbf{a} \equiv (a_1, ..., a_m)$, $\mathbf{b} \equiv (b_1, ..., b_m)$ and $\mathbf{c} \equiv (c_1, ..., c_m)$ are three sequences of $m$ real numbers, $\mathbf{a}$ being a sequence for which only the ordering of values is important. The sequences $\mathbf{a}$ and $\mathbf{b}$ will later be identified with, respectively, the set of dissimilarities

$(d_{ij})$ and the set of interpoint distances $(\hat{d}_{ij})$ within a configuration of points; c is a sequence used in the comparison of a and b.

Two possible definitions of monotonicity are:

(i) c is primarily monotone increasing (PMI) over a if

$$a_r < a_s \ \Rightarrow\ c_r \leq c_s \ (1 \leq r, s \leq m); \qquad (6.12)$$

(ii) c is secondarily monotone increasing (SMI) over a if

$$a_r \leq a_s \ \Rightarrow\ c_r \leq c_s \ (1 \leq r, s \leq m). \qquad (6.13)$$

These two definitions of monotonicity differ only in their treatment of tied values in the sequence a. In the secondary definition, these tied values must be preserved in c: if $a_r$ equals $a_s$, then $c_r$ must equal $c_s$. In the primary definition of monotonicity, ties in a may be broken in either direction in c.

The sequence c is required to resemble b as closely as possible, subject to being monotone (either PMI or SMI) over a. As an example, consider the sequences a = (1, 2, 4, 4, 6, 8, 9, 10, 11, 15) and b = (1, 4, 5, 6, 7, 8, 12, 13, 13, 14). The points $\{(a_k, b_k), k = 1, ..., 10)\}$ are plotted as crosses in Fig. 6.5. If the monotonicity requirement is PMI, the equality between $a_3$ and $a_4$ may be broken, and by choosing $c_k = b_k (k = 1, ..., 10)$ one can ensure a perfect resemblance between c and b, with c satisfying the primary monotonicity condition defined in (6.12). If the monotonicity requirement is SMI, $c_3$ must equal $c_4$, and because $b_3$ does not equal $b_4$ it is not possible to have c = b. In order to specify the optimal shared value for $c_3$ and $c_4$, it is necessary to know how departure from a perfect fit between b and c is measured. Kruskal (1964a) suggested that a sum of squares criterion be used:

$$S^*(\mathbf{c}) \ \equiv\ \Sigma_{k=1}^m (b_k - c_k)^2.$$

This criterion is minimized when $c_3$ and $c_4$ are both chosen to be $5\frac{1}{2}$, the mean of $b_3$ and $b_4$. More generally, c is split up into a set of blocks of elements with consecutive subscripts, e.g. $(c_r, c_{r+1}, ..., c_s)$, such that each element in the block equals the mean of the corresponding set of values in b $(\Sigma_{k=r}^s b_k/(s - r + 1))$, and such that the common values increase with the subscripts.

For given sequences a and b, the sequence c which reduces $S^*(\mathbf{c})$ to its minimum value ($S^*$, say), subject to being PMI (SMI) over a, is called the least squares monotone regression of b on a: $S^*$ is called the primary (secondary) raw stress. The secondary least

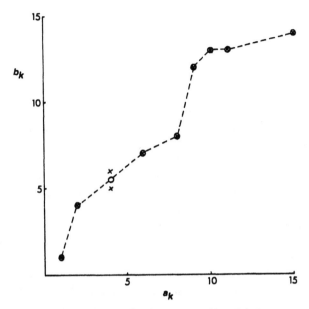

Figure 6.5 *A plot of the artificial sequence* $\{(a_k, b_k), k = 1, ..., 10\}$ *described in the text, and the secondary least squares monotone regression* **c** *of* **b** *on* **a**: *the points* $(a_k, b_k)$ *and* $(a_k, c_k)$ *are represented by crosses and open circles, respectively. The regression has raw stress* $S^* = \frac{1}{2}$.

squares monotone regression **c** of **b** on **a** for the artificial example is shown by the set of circles in Fig. 6.5. For illustrative purposes, dashed straight lines link the circles in Fig. 6.5, but the form of the regression function between neighbouring circles is required only to be monotone.

This methodology may be applied to the comparison of sets of dissimilarities $(d_{ij})$ and distances $(\hat{d}_{ij})$: for example, the $m$ $(= 10)$ elements of **a** could be the $n(n-1)/2$ $(= 10$ for $n = 5)$ pairwise dissimilarities within a set of five objects, and the elements of **b** could be the interpoint distances within a set of five points representing the same five objects. The interpoint distance is usually defined to be Euclidean distance, but other Minkowski metrics can also be used: Arabie (1991) discusses use of the city block metric in this context.

The raw stress $S^*$ provides a measure of the resemblance between given sets of dissimilarities and distances which is not invariant under uniform dilation of the configuration of points. This undesirable property is removed by dividing by a normalizing factor, e.g. $T^*$ $\equiv \sum \hat{d}_{ij}^2$, and the normalized stress, $S$, is defined by:

$$S \equiv \sqrt{\frac{S^*}{T^*}} = \left[\frac{\sum(\hat{d}_{ij} - c_{ij})^2}{\sum \hat{d}_{ij}^2}\right]^{1/2}, \qquad (6.14)$$

thus ensuring that $S$ lies between 0 and 1; lower values indicate better fits. In this expression, $(c_{ij})$ is the (primary/secondary) least squares monotone regression of $(\hat{d}_{ij})$ on $(d_{ij})$ and the summations are taken over some or all of the pairs $(i, j)$.

Thus far, it has been assumed that the set of interpoint distances $(\hat{d}_{ij})$ is known. In practice, a set of dissimilarities $(d_{ij})$ is given, from which one wishes to obtain a configuration of points in $t$ (say) dimensions, with coordinate values $\{x_{ik} \ (i = 1, ..., n; k = 1, ..., t)\}$, whose interpoint distances $(\hat{d}_{ij})$ lead to the smallest possible value of $S$, i.e. $S$ is a function of the $nt$ variables $(x_{ik})$. A solution is found by successively improving an initial configuration using an iterative function-minimization algorithm (e.g. Kruskal, 1964b). The danger that the final configuration obtained is a poor locally optimal solution which is far removed from the globally optimal solution can be reduced by repeating the analysis several times, starting from different initial configurations of points.

The secondary definition of monotonicity, specified in (6.13), appears to be less satisfactory than the primary definition, often leading to a configuration that is less readily interpretable and/or one that has become trapped in a poor local minimum of the stress function (Sibson, 1972; Lingoes and Roskam, 1973).

The extent of the summations in Expression (6.14) has not yet been specified. Usually, one would sum over all $n(n - 1)/2$ pairs $(i, j)$ $(1 \leq j < i \leq n)$, but other definitions are possible. Thus, the set of pairs could be divided into subsets, with separate monotone regressions being carried out within each subset and the lack of fit within each subset being combined into a single measure of stress: for example, it may be relevant to carry out comparisons only within each row of a dissimilarity matrix if the objects have been ranked with respect to their similarity to each object (Sibson, 1972).

Missing values can also be handled within this framework: if the dissimilarities $\{d_{i(r)}, d_{j(r)} \ (r = 1, ..., s)\}$ are not available, these pairs of subscripts would not be included in the summation when the stress is evaluated. This feature is particularly relevant if the dissimilarities have to be obtained from pairwise assessments of the objects made by an observer, because such an individual may find it difficult to provide consistent assessments over a large number of comparisons. Several simulation studies have investigated the extent to which known configurations of points were retrieved when some of the values in the corresponding dissimilarity matrices were suppressed and non-metric multidimensional scaling was used to analyse the incomplete dissimilarity matrices. Spence and Domoney (1974) found that the configurations were most closely recovered if the missing values were chosen so as to leave a cyclic pattern, but that the results were not greatly inferior to this if the missing values were chosen at random; one-third of the dissimilarities could be suppressed without markedly affecting a configuration. Graef and Spence (1979) divided the dissimilarities into three groups, comprising large, medium and small values, and reconstructed the configuration using only two of these groups, finding that the configuration was badly distorted when the larger dissimilarities were discarded, but not greatly affected by discarding either of the other two groups of dissimilarities. Interactive procedures that sequentially select pairs of objects to be compared have also been proposed: Cliff et al. (1977) and Green and Bentler (1979) describe relevant algorithms for use in obtaining classical scaling solutions.

It is necessary to specify an appropriate value for $t$, the number of dimensions in which the configuration of points is to be obtained. There is no direct relationship between configurations in $t_1$ and $t_2$ dimensions (where $t_1 < t_2$), as there is for the results of a principal coordinates analysis, in which the lower-dimensional representation is simply a projection onto the first $t_1$ axes of the $t_2$-dimensional solution. With non-metric multidimensional scaling, the representation in each number of dimensions has to be obtained separately, although one solution might be of use in providing an initial configuration in a smaller number of dimensions. It is common to obtain configurations in $t$ dimensions for a small range of values of $t$ and to examine a plot of the stress $S(t)$ against the number of dimensions. An 'elbow' in this stress function, such as is shown in Fig. 6.6, would suggest when little was to be gained by obtaining a higher-dimensional representation. Often, however, no clear elbow is visible in such plots. An alternative criterion is based on the interpretability of the output (Kruskal, 1964a):

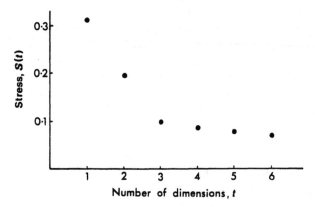

Figure 6.6  *A plot of the stress, $S(t)$, against the number of dimensions, t, of the configuration of points. There is little reduction to be obtained in the stress by scaling in more than three dimensions.*

'if the $t$-dimensional solution provides a satisfying interpretation, but the $(t + 1)$-dimensional solution reveals no further structure, it may be well to use only the $t$-dimensional solution.'

More formal approaches to interpreting the value of the stress and assessing appropriate values for $t$ are discussed by Cox and Cox (1994, Sections 3.4, 3.5). The presentation in this book emphasises the informal assessment of configurations of points, and hence concentrates on two-dimensional configurations.

Fig. 6.7 shows a two-dimensional representation of the acoustic confusion data provided by non-metric multidimensional scaling, for which the stress takes the value 0.133. A plot of the interpoint distances against the original pairwise similarities is given in Fig. 6.8, confirming the earlier result that the relationship between them is not straightforward.

Some investigators are often concerned to identify interesting axes or directions in the graphical representation that they have obtained. The results of a non-metric multidimensional scaling are invariant under rotations/reflections, and hence the coordinate axes are indeterminate in this respect. The configuration of points can be rotated to obtain principal component axes, but these axes need not have a useful interpretation. Sometimes, there are readily interpretable linear trends in the graphical representa-

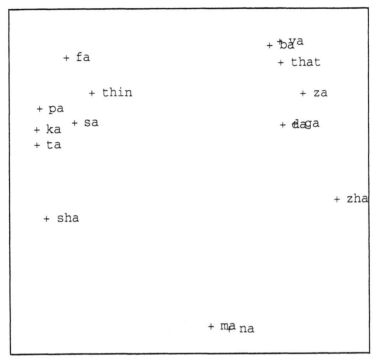

Figure 6.7 *A non-metric multidimensional scaling of the acoustic confusion data.*

tion: for example in Morse code data analysed by Shepard (1963), the two-dimensional plot of 36 Morse code symbols showed two nearly linear and orthogonal axes, which could be interpreted as the number of components in the symbol and the relative frequencies of dots and dashes in the symbol. On other occasions, however, such linear trends were not readily discernible: for example, in analyses of the perceived similarities of different colours (Shepard, 1962b) and colour names (Rapoport and Fillenbaum, 1972), the two-dimensional plots showed the colours situated roughly around the circumference of a circle. As far as the assessment of graphical representations is concerned, the implication is that the naming of axes or interpretation of directions in space should be conducted with caution.

It can also be of interest to link variables (which might or might

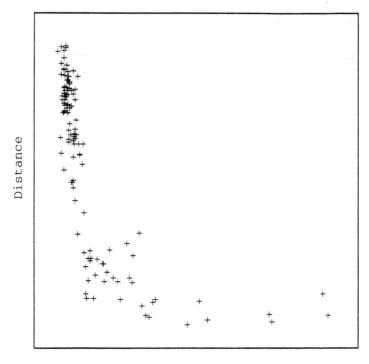

Figure 6.8 *A plot of the interpoint distances in Fig. 6.7 against the pairwise similarities for the acoustic confusion data.*

not have contributed to the measure of pairwise dissimilarity between the objects) with directions in graphical representations. The relationship of a variable **y** to the set of objects can be indicated by using multiple linear regression (with **y** as the dependent variable and the coordinate values as the independent variables) to identify a direction in space which reflects the values of **y**; if a model for a two-dimensional representation provides a reasonable fit, a line can be inserted into the plot (Kruskal and Wish, 1978). Alternative ways of obtaining graphical representations that provide information about variables as well as objects are presented in Section 6.5.

If there is an upper limit to the value that dissimilarities can take, two-dimensional representations of data that are believed to depend on a single trend often show a configuration of points resembling a horseshoe. This can be explained by realising that objects that are more than a certain threshold value apart in terms of the trend are all likely to have their pairwise dissimilarities close to this largest value. Modified (dis)similarity coefficients intended to counter this effect are presented by Kendall (1971) and Williamson (1978).

It is not advisable to use the methodology described in this section to obtain a one-dimensional representation (or seriation) of a set of objects: Hubert and Arabie (1986, 1988) show that gradient-based methods of optimization can fail in this instance, and describe an alternative dynamic programming algorithm; other algorithms for one-dimensional multidimensional scaling have been proposed by Pliner (1996) and Lau, Leung and Tse (1998).

Much attention has been paid in recent years to more formal modelling and assessment in multidimensional scaling, as well as to the development of more general multidimensional scaling models. Some investigators have questioned the extent to which it is useful to incorporate more formal modelling and testing within the multidimensional scaling paradigm, preferring to see the methodology purely as a tool for exploratory data analysis, a view expressed by some of the discussants of Ramsay (1982). Although this book concentrates on describing exploratory aspects of multidimensional scaling, it seems useful to have the possibility of undertaking more detailed analyses of relevant data sets. A review of such work is provided by Carroll and Arabie (1997), who interpret the words 'multidimensional scaling' in a wider sense than is done here.

This section is completed with a comparison of the methodology described in the last two sections. A brief summary of some of the assumptions and properties of principal coordinates analysis and non-metric multidimensional scaling is given in Table 6.1. There have been investigations that compare the analyses of the same data sets provided by several different graphical procedures. For example, Sibson, Bowyer and Osmond (1981) discuss several possible models for obtaining a set of dissimilarities from an underlying configuration of points, and describe a simulation study that assessed their ability to retrieve the original configuration, reporting that – provided the iterations started from a reasonable configuration – non-metric multidimensional scaling was never significantly worse, and under some models for the dissimilarities was considerably better, than principal coordinates analysis and least squares

Table 6.1 *A comparison of the methods of principal coordinates analysis and non-metric multidimensional scaling.*

| Principal coordinates analysis | Non-metric multidimensional scaling |
| --- | --- |
| It is assumed that the pairwise dissimilarities are squared distances between pairs of points in a graphical representation | It is assumed that the dissimilarities and distances are monotonically related, and that the dissimilarities contain no information beyond their rank ordering |
| A $t$-dimensional representation can be obtained simultaneously for all values of $t$ for which the corresponding eigenvalues of the inner product matrix are positive. The solution in $t_1$ dimensions is obtained from the solution in $t_2$ dimensions $(t_1 < t_2)$ by projecting the points onto the first $t_1$ principal axes | A solution has to be obtained for each number of dimensions separately. An iterative function-minimization algorithm has to be used, with no guarantee that the global minimum solution will be obtained. There need be no simple relationship between configurations in different numbers of dimensions |
| The method makes less heavy demands on computing resources | The method makes heavy demands on computing resources |

scaling methods which seek to minimize Expression (6.11). Rowell, McBride and Palmer (1973) compared the output from several clustering and graphical methods of analysis, noting that, in general, clustering algorithms induced a larger distortion in the larger pairwise dissimilarities, principal coordinates analysis induced a larger distortion in the smaller dissimilarities, and non-metric multidimensional scaling spread out the distortion more evenly over the complete set of dissimilarities. As one would expect, principal coordinates analysis and least squares scaling are less robust than non-metric multidimensional scaling to the presence of gross errors in the data (Spence and Lewandowsky, 1989).

Just as with clustering algorithms, it would be valuable to have guidelines indicating relevant graphical methods of analysis for data sets under investigation, but it is rarely the case that the background to the problem indicates that one method of analysis is peculiarly appropriate for a given study. Non-metric multidimensional scaling has the advantage of making minimal assumptions about how distances are related to dissimilarities, but both it and least squares scaling make heavy demands on computing resources, requiring the minimization of a function of $nt$ variables for each value of the number of dimensions, $t$. By contrast, the main computational burden in principal coordinates analysis involves the eigenanalysis of an $(n \times n)$ matrix. Given this fact, particularly for larger data sets it would seem preferable initially to use principal coordinates analysis (or some of the methodology described in the next section). The adequacy of the resulting configuration can then be assessed, with the assistance of Expression (6.10) and a plot of the $n(n-1)/2$ values of $(d_{ij}, \hat{d}_{ij})$, such as is shown in Fig. 6.4. One could then consider employing other graphical methods, possibly using the results of principal coordinates analysis to provide an initial configuration of points.

More general multidimensional scaling models, relevant for the analysis of higher-way and higher-mode data matrices, are described by Arabie, Carroll and DeSarbo (1987) and Cox and Cox (1994).

## 6.4 Interactive graphics and self-organizing maps

This section gathers together a miscellaneous collection of methodology for obtaining and assessing graphical representations of a set of objects.

### 6.4.1 Interactive graphics

The results of analysing data using the methodology described thus far in the chapter have been viewed in a *static* context, for example as a configuration of labelled points printed on paper. However, computing developments have allowed investigators to explore features of their data more interactively; thus, information extracted from the data can influence the direction of further analyses, which can be immediately carried out.

Much of this work has been concerned with enabling investigators to manipulate high-dimensional configurations of points. The pioneer system PRIM-9, developed in the early 1970's (Fisherkeller, Friedman and Tukey, 1988), provided the facilities 'picturing', 'rotation', 'isolation' and 'masking' (whose initial letters spell out the acronym). Thus, the configuration of points can be rotated in any direction and projections onto two-dimensional subspaces can be viewed; in addition, attention can be restricted to specified subsets of objects or regions of space. The projections that are displayed can be left to the investigator to specify, or an automated method of selection can be used. Asimov (1985) describes ways of selecting a sequence of projections such that any possible projection is sufficiently close to one of the members of the sequence. 'Projection pursuit' methods (Kruskal, 1969, 1972; Friedman and Tukey, 1974; Huber, 1985; Jones and Sibson, 1987) seek projections of the data (generally onto one or two dimensions) that optimize an index of 'interestingness'. For example, Kruskal (1972) investigates a standardized version of the coefficient of variation of the $\frac{1}{4}$th power of the interpoint distances, whereas Friedman and Tukey's (1974) index is a product of a trimmed standard deviation and a measure of the local density of points. Krzanowski and Marriott (1994, Section 4.13) and Ripley (1996, Section 9.1) provide reviews of many other indices that have been proposed; in a classification context, indices that find regions with a locally high density of points are particularly of interest. Clusters of points identified in a projection would be removed and the analysis repeated on the reduced data set.

Other developments in interactive graphics have included the option of 'brushing', whereby objects chosen in one display are highlighted in another display (Becker and Cleveland, 1987). Fig. 6.9 presents a scatterplot matrix (Tukey and Tukey, 1981a) of the four-dimensional principal coordinates analysis representation of the acoustic confusion data, comprising plots of the objects on all pairs of the first four principal axes. The objects have been given the same labels in each of the plots, allowing one to synthesize information from different plots. However, such a synthesis would in general be difficult for larger data sets, for which brushing provides an effective alternative.

When a configuration is rotated at a suitable speed, the impres-

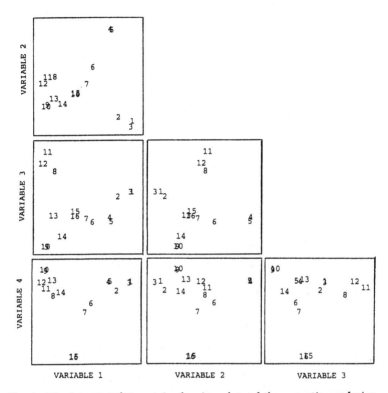

Figure 6.9 *A scatterplot matrix showing plots of the acoustic confusion data on each pair of principal axes selected from the four-dimensional principal coordinates analysis of the data.*

sion of three-dimensional data can be provided, an effect that can be enhanced by adding stereo vision (Papathomas, Schiavone and Julesz, 1987; Becker, Cleveland and Weil, 1988).

Further comments on interactive graphics methodology and software are given in Cleveland and McGill (1988) and Young, Faldowski and McFarlane (1993).

### 6.4.2 Self-organizing maps

Kohonen (1982, 1990) developed methods of transforming a set of $p$-variate observations into a set ($A$, say) which has some imposed structure; for example, $A$ could comprise $m$ points arranged on a regular two-dimensional grid. Associated with the $j$th point in the grid is a $p$-dimensional vector $\mathbf{y_j}$ $(j = 1, ..., m)$, whose initial values are chosen randomly. Each of the input observations is compared with the set of vectors in order to identify one to which it is closest, and the vectors are iteratively updated: thus, if $\mathbf{y_k}$ is a vector that is closest to the input observation $\mathbf{x}$,

$$\mathbf{y_j} \leftarrow \mathbf{y_j} + \alpha(\mathbf{x} - \mathbf{y_j}) \text{ for all } \mathbf{y_j} \in N_k,$$

where $N_k$ is the set of points belonging to the neighbourhood of the $k$th point in the grid, and $0 < \alpha < 1$. The definitions of both the neighbourhood and $\alpha$ are altered (reduced in size) as the iterations proceed. The input observations are iteratively presented for comparison with the vectors until convergence is achieved.

A plot of the configuration of points, indicating to which of the points in the grid each of the input observations is assigned, is called a 'self-organizing feature map' or 'topological map'.

As an illustration, this methodology was used to obtain a topological map of the combined Abernethy Forest data sets on a (20 × 20) square grid. The results are shown in Table 6.2, in which each grid point is either labelled with one of the input observations assigned to it, or left blank if it has no input observations. Six of the 90 input observations are not shown because they were overwritten by other inputs assigned to the same grid point; thus, the following pairs of input observations are located at the same grid point: (027, 028), (423, 424), (428, 429), (435, 437), (445, 039), (448, 449). The graphical representation shown in Table 6.2 is in good agreement with the analyses reported in Section 3.2.3.

By decreasing the number of grid points, a greater amount of clustering of input vectors can be achieved. If the neighbourhood of each grid point always contains only the grid point itself, there is no linkage between neighbouring points in the grid and the process reduces to a $k$-means type of clustering (Ripley, 1996, Section 9.4). Alternatively, the $m$ classes of input vectors in the topological map can be partitioned into a smaller number of classes by a constrained clustering algorithm (see Section 5.2) that takes into account the spatial contiguity of the grid points (Murtagh, 1995). Further links between Kohonen's self-organizing map method and other data analytic methods are discussed by Murtagh and Hernández-Pajares (1995) and Bock (1998).

The self-organizing map method is usually described in the terminology of artificial neural networks. This topic comprises a mis-

Table 6.2 *Self-organizing feature map of the combined Abernethy Forest data sets. Grid points associated with input vectors are labelled by three numbers, the first number indicating the data set to which the input vector belongs: '4' denotes the 1974 data set, '0' denotes the 1970 data set.*

| 449 | 434 | 437 |     | 440 | 438 | 008 | 003 | 004 | 006 | 007 | 016 | 011 |     | 405 |
|-----|-----|-----|-----|-----|-----|-----|-----|-----|-----|-----|-----|-----|-----|-----|
|     |     |     | 441 |     |     |     |     |     |     |     |     |     | 403 | 406 |
| 447 | 446 | 436 | 439 |     | 041 | 021 | 005 | 012 |     | 010 | 014 | 013 | 408 | 407 |
| 039 | 038 | 035 |     | 036 |     |     | 002 |     | 001 |     |     | 023 |     | 404 |
| 444 |     | 040 |     |     | 022 | 034 | 020 | 018 | 019 |     | 015 | 017 | 401 | 009 |
|     | 443 |     |     |     |     |     |     |     |     | 024 |     |     |     | 402 |
| 433 | 442 | 033 |     |     |     |     | 025 |     |     |     |     |     | 409 | 410 |
| 432 |     | 032 | 031 | 029 |     | 030 |     |     |     |     |     |     |     | 412 |
|     | 431 |     |     |     |     |     |     |     |     |     |     |     |     | 413 |
| 430 |     | 031 | 418 |     | 417 | 034 |     |     |     |     |     |     |     | 411 |
| 429 | 425 |     | 416 |     |     |     |     |     |     |     |     |     |     |     |
|     | 427 |     | 422 | 420 | 419 | 028 | 026 |     |     |     |     |     | 415 | 414 |
| 426 | 424 |     | 421 |     |     |     |     |     |     |     |     |     |     |     |

cellaneous collection of methods that use networks to organise the information contained in data; noteworthy features are that algorithms can usually be implemented in parallel and that large data sets can be analysed. Much neural network methodology has addressed the problem of assigning objects to existing classes (Bishop, 1995; Ripley, 1996), but there has also been some 'unsupervised' neural network methodology (of which self-organizing map methods are one example), in which no external information is provided about any classes to which objects may belong. Such networks have been used in the identification of directions in projection pursuit, with algorithms implemented in parallel (Intrator, 1992; Intrator and Cooper, 1992). Other areas of application are discussed by Mao and Jain (1995), Murtagh (1996) and Bock (1998).

## 6.5 Biplots

The methodology described thus far in the chapter yields low-dimensional configurations of points representing a set of objects. It can also be used to provide configurations of points representing a set of variables (e.g. by obtaining a multidimensional scaling of a matrix containing the pairwise dissimilarities within a set of variables). However, in each case the main output is a graphical representation of a single set of 'entities'. As described in Section 6.3, information about the variables that describe a set of objects can subsequently be incorporated into a configuration of points representing the objects. This section describes more direct ways of providing simultaneous representations both of a set of objects and of the set of variables describing the objects. Thus, each object and each variable is represented by a point or a vector in the joint representation, which is called a *biplot*. Such biplots allow one to assess

(i) similarities within the set of objects

(ii) similarities within the set of variables

(iii) links between objects and variables.

The following two subsections describe basic biplots for the situations in which the variables describing the objects are, respectively, all quantitative and all categorical. A fuller description of the subject is given by Gower and Hand (1996).

### 6.5.1 Principal component biplots

This method of analysis was proposed by Gabriel (1971). The pattern matrix $\mathbf{X} \equiv (x_{ik})$ is an $(n \times p)$ matrix describing $n$ objects in terms of $p$ quantitative variables, and is assumed to have been standardized so that the mean on each variable is zero. If $\mathbf{X}$ is of rank $r$, it can be expressed as the product of an $(n \times r)$ matrix $\mathbf{G}$ and an $(r \times p)$ matrix $\mathbf{H}'$:

$$\mathbf{X} = \mathbf{GH}'. \qquad (6.15)$$

Equivalently,

$$x_{ik} = \mathbf{g}_i'\mathbf{h_k} \ (i = 1, ..., n; k = 1, ..., p), \qquad (6.16)$$

where $\mathbf{g}_i'$ denotes the $i$th row of $\mathbf{G}$, and $\mathbf{h_k}$ denotes the $k$th column of $\mathbf{H}'$. In other words, the vector $\mathbf{g_i}$ is associated with the $i$th row of $\mathbf{X}$, the vector $\mathbf{h_k}$ is associated with the $k$th column of $\mathbf{X}$, and the set $\{\mathbf{g_1}, ..., \mathbf{g_n}, \mathbf{h_1}, ..., \mathbf{h_p}\}$ provides a representation of $\mathbf{X}$ as $(n+p)$ vectors in $r$-dimensional space. However, Equation (6.15) does not provide a unique characterization of a biplot: if $\mathbf{R}$ is any $(r \times r)$ non-singular matrix, an alternative representation is provided by $\mathbf{X} = \tilde{\mathbf{G}}\tilde{\mathbf{H}}'$, for $\tilde{\mathbf{G}} \equiv \mathbf{GR}'$ and $\tilde{\mathbf{H}} \equiv \mathbf{HR}^{-1}$.

Biplots are usually depicted in two dimensions, and the following description concentrates on this case. If $r = 2$, the two-dimensional representation captures all the information in $\mathbf{X}$. If $r > 2$, one can obtain a matrix $\mathbf{X_2}$ which is the rank-2 least squares approximation of $\mathbf{X}$, and the biplot of $\mathbf{X_2}$ provides an approximation to the biplot of $\mathbf{X}$.

Suppose that the singular value decomposition of $\mathbf{X}$ is

$$\mathbf{X} = \mathbf{T\Theta U}' = \Sigma_{j=1}^{r}\theta_{jj}\mathbf{t_j}\mathbf{u_j'}, \qquad (6.17)$$

where $\mathbf{T} \equiv (\mathbf{t_1}, ..., \mathbf{t_n})$ and $\mathbf{U} \equiv (\mathbf{u_1}, ..., \mathbf{u_p})$ are orthogonal matrices, and $\mathbf{\Theta}$ is a diagonal $(n \times p)$ matrix whose only non-zero elements are $\{\theta_{jj} \ (j = 1, ..., r)\}$, where $r \leq min(n - 1, p)$, and

$$\theta_{11} \geq \theta_{22} \geq ... \geq \theta_{rr} > 0.$$

As described in Section 6.2, a rank-2 least squares approximation of $\mathbf{X}$ is

$$\mathbf{X_2} \equiv \mathbf{T_2\Theta_2U_2'} = (\mathbf{t_1}, \mathbf{t_2}) \begin{pmatrix} \theta_{11} & 0 \\ 0 & \theta_{22} \end{pmatrix} \begin{pmatrix} \mathbf{u_1'} \\ \mathbf{u_2'} \end{pmatrix}. \qquad (6.18)$$

Because $\mathbf{T}_2$ and $\mathbf{U}_2$ are obtained from the first two columns of the orthogonal matrices $\mathbf{T}$ and $\mathbf{U}$,

$$\mathbf{T}_2'\mathbf{T}_2 = \mathbf{I}_2 = \mathbf{U}_2'\mathbf{U}_2,$$

where $\mathbf{I}_2$ is the $(2 \times 2)$ identity matrix.

It is conventional to replace each vector $\mathbf{g}_i$ by a point located at the end of the vector. The indeterminacy in the definitions of $\mathbf{G}$ and $\mathbf{H}$, mentioned above, can be removed by requiring the distance between pairs of g-points in a two-dimensional graphical representation to equal the distance (in a suitable metric) between the corresponding pairs of objects defined in the matrix $\mathbf{X}_2$. Thus, $\mathbf{G}$ is defined by

$$\mathbf{GG}' = \mathbf{X}_2\mathbf{M}\mathbf{X}_2',$$

where $\mathbf{M}$ specifies the metric used on the rows of $\mathbf{X}_2$. Two special cases are:

   (i) $\mathbf{M} = \mathbf{I}_2$
   (ii) $\mathbf{M} = \mathbf{S}_2^{-1}$, where $\mathbf{S}_2 \equiv \mathbf{X}_2'\mathbf{X}_2/(n-1)$, the sample covariance matrix of $\mathbf{X}_2$.

These definitions ensure that Euclidean distance between each pair of g-points corresponds to ($i$) Euclidean distance, ($ii$) Mahalanobis distance, between the corresponding rows of $\mathbf{X}_2$, as shown below.

   ($i$) From Equation (6.18),

$$\mathbf{X}_2\mathbf{X}_2' = (\mathbf{T}_2\boldsymbol{\Theta}_2\mathbf{U}_2')(\mathbf{U}_2\boldsymbol{\Theta}_2\mathbf{T}_2') = (\mathbf{T}_2\boldsymbol{\Theta}_2)(\mathbf{T}_2\boldsymbol{\Theta}_2)'.$$

Hence, choosing

$$\mathbf{G} = \mathbf{T}_2\boldsymbol{\Theta}_2 \tag{6.19}$$
$$\mathbf{H} = \mathbf{U}_2 \tag{6.20}$$

ensures that $\mathbf{GG}' = \mathbf{X}_2\mathbf{X}_2'$. From the theory described in Section 6.2, the matrix $\mathbf{G}$ provides a principal components analysis of $\mathbf{X}$.

   ($ii$) From Equation (6.18), the sample covariance matrix of $\mathbf{X}_2$ is given by

$$(n-1)\mathbf{S}_2 = (\mathbf{T}_2\boldsymbol{\Theta}_2\mathbf{U}_2')'(\mathbf{T}_2\boldsymbol{\Theta}_2\mathbf{U}_2') = \mathbf{U}_2\boldsymbol{\Theta}_2^2\mathbf{U}_2'. \tag{6.21}$$

Hence,

$$\mathbf{X_2 S_2^{-1} X_2'} = (n-1)(\mathbf{T_2 \Theta_2 U_2'})(\mathbf{U_2 \Theta_2^{-2} U_2'})(\mathbf{U_2 \Theta_2 T_2'})$$
$$= (n-1)\mathbf{T_2 T_2'}.$$

Choosing

$$\mathbf{G} = \sqrt{(n-1)}\mathbf{T_2} \qquad\qquad (6.22)$$
$$\mathbf{H} = \mathbf{U_2 \Theta_2'}/\sqrt{(n-1)} \qquad\qquad (6.23)$$

ensures that $\mathbf{GG'} = \mathbf{X_2 S_2^{-1} X_2'}$. (Gabriel's (1971) description replaces $(n-1)$ in Equations (6.22) and (6.23) by $n$ because of a different definition of the sample covariance matrix.)

From Equations (6.21) and (6.23), $\mathbf{S_2} = \mathbf{HH'}$, hence information about the sample covariance matrix is provided by the set of h-vectors; in particular,

- the standard deviation of the $k$th variable is given by the length of $\mathbf{h_k}$

- the correlation between two variables is given by the cosine of the angle between the corresponding pair of h-vectors.

These results are exact for the analysis of $\mathbf{X_2}$. However, if $\mathbf{X_2}$ does not capture most of the variability in $\mathbf{X}$, the two-dimensional biplot may not provide a good approximation of the relationships in $\mathbf{X}$. The accuracy with which the elements of $\mathbf{X}$ have been plotted can be measured by

$$\Sigma_{i=1}^2 \theta_{ii}^2 / \Sigma_{j=1}^r \theta_{jj}^2 \qquad\qquad (6.24)$$

(compare Equation (6.10)). However, the adequacy of different aspects of the biplot differs: for example (Gabriel, 1971), the accuracy with which the sample covariance matrix is portrayed in the biplot based on the Mahalanobis distance option is

$$\Sigma_{i=1}^2 \theta_{ii}^4 / \Sigma_{j=1}^r \theta_{jj}^4. \qquad\qquad (6.25)$$

As an illustration, consider the following pattern matrix, in which five objects are described by three quantitative variables:

$$
\mathbf{X} = \begin{pmatrix} 60 & 80 & -240 \\ -213 & 66 & 180 \\ 123 & -186 & 180 \\ -9 & 38 & -60 \\ 39 & 2 & -60 \end{pmatrix}.
$$

This matrix is of rank 2, with singular value decomposition

$$
\mathbf{X} = \mathbf{T_2 \Theta_2 U_2'}
$$

$$
= \begin{pmatrix} 2/3 & 0 \\ -1/2 & 7/10 \\ -1/2 & -7/10 \\ 1/6 & 1/10 \\ 1/6 & -1/10 \end{pmatrix} \begin{pmatrix} 390 & 0 \\ 0 & 300 \end{pmatrix} \begin{pmatrix} 3/13 & 4/13 & -12/13 \\ -4/5 & 3/5 & 0 \end{pmatrix}.
$$

Using Equations (6.22) and (6.23), the Euclidean distance between pairs of g-points can be made equal to the Mahalanobis distance between the corresponding pairs of rows of $\mathbf{X}$ by defining

$$
\mathbf{G} = \sqrt{(n-1)}\mathbf{T_2} = \begin{pmatrix} 4/3 & 0 \\ -1 & 7/5 \\ -1 & -7/5 \\ 1/3 & 1/5 \\ 1/3 & -1/5 \end{pmatrix}
$$

$$
\mathbf{H} = \mathbf{U_2 \Theta_2'}/\sqrt{(n-1)} = \begin{pmatrix} 45 & -120 \\ 60 & 90 \\ -180 & 0 \end{pmatrix}.
$$

This biplot is plotted in Fig. 6.10, in which the h-vectors have been scaled down by a factor of 100 and labelled $V_1 - V_3$. The relative directions of these vectors shows that the three variables are negatively correlated with one another. The fact that $V_3$ is the longest vector indicates that the third variable has the most heterogeneous set of values. Each of the five g-points has been projected onto $V_2$, whose extension in the negative direction is shown by a dashed line. The relative positions of the projected values perfectly reproduce the values in the second column of $\mathbf{X}$. If desired, the h-vectors can be calibrated, so that the predicted values could be read off directly from the Figure.

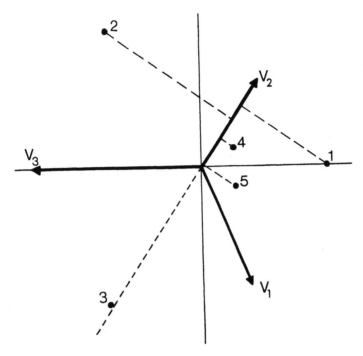

Figure 6.10  *A biplot of the artificial data set described in the text.*

Fig. 6.11 shows a biplot of the combined Abernethy Forest data sets, in which the Mahalanobis distance option has been chosen; the h-vectors have been scaled up by a factor of 10. The configuration of points in this Figure differs from the principal components analysis of the same data set shown in Fig. 3.3 only by having the vertical axis stretched relative to the horizontal axis. The directions of the h-vectors are informative. For example, the Figure indicates that the topmost (latest) samples in each core (see Table 3.4) tend to have above average values of the second variable (*Pinus* pollen), the next samples down the cores tend to have above average values of the first (*Betula*) and third (*Corylus/Myrica*) variables, and the earliest samples tend to have above average values of the last five variables (mainly grasses and sedges).

Extensions to the methodology described in this section include a weighted least squares approximation of $X_2$, which can be used to obtain biplots when some of the entries in a pattern matrix

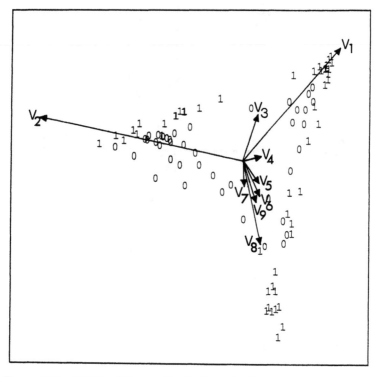

Figure 6.11 *A biplot of the combined Abernethy Forest data sets, showing the h-vectors associated with each of the nine species of pollen; samples from the 1970 data set are coded '0', and samples from the 1974 data set are coded '1'.*

are missing (Gabriel and Zamir, 1979), and a dynamic biplot for monitoring over time whether or not there are any changes in a multivariate process (Sparks, Adolphson and Phatak, 1997).

### 6.5.2 Multiple correspondence analysis

The method of multiple correspondence analysis represents a set of objects and the levels of a set of categorical variables that describe the objects as a configuration of points in a single graphical representation. There have been several different derivations of multiple correspondence analysis: for example, it has been regarded as a

Table 6.3 *Description of ten objects in terms of three categorical variables.*

| Object | Gender | Hair colour | Eye colour |
|--------|--------|-------------|------------|
| 1      | male   | fair        | blue       |
| 2      | male   | black       | brown      |
| 3      | female | fair        | green      |
| 4      | female | fair        | blue       |
| 5      | male   | brown       | brown      |
| 6      | male   | black       | green      |
| 7      | female | brown       | green      |
| 8      | female | brown       | brown      |
| 9      | male   | fair        | blue       |
| 10     | male   | brown       | blue       |

way of assigning numerical values to category levels (Gifi, 1990, Chapter 3), and as a generalization of (classical) correspondence analysis of two-way contingency tables (Greenacre, 1984, Chapter 5). This account follows the approach described in Gower and Hand (1996), in which the results are obtained from a principal components analysis of a suitably-defined matrix.

Consider the data presented in Table 6.3, in which ten individuals are described by three categorical variables: gender {male, female}, hair colour {fair, brown, black}, and eye colour {blue, green, brown}.

These data can alternatively be described by a $(10 \times 8)$ indicator matrix, $\mathbf{Y}$, in which each category level is given a separate column and an entry of 1 indicates the relevant level of the category. Thus, the data given in Table 6.3 can also be described as shown in Table 6.4.

The analysis is carried out on the matrix

$$\mathbf{X} \equiv p^{-1/2}\mathbf{Y}\mathbf{C}^{-1/2}, \qquad (6.26)$$

where $\mathbf{C}$ is a diagonal matrix whose $j$th diagonal entry holds the $j$th column sum of $\mathbf{Y}$, and $p$ denotes the number of categorical variables. It is assumed in the following presentation that $\mathbf{X}$ has then been standardized to have all its columns summing to zero (although this standardization could alternatively be obtained by discarding a trivial dimension in the singular value decomposition).

Table 6.4 *Indicator matrix of the data given in Table 6.3.*

| Gender | | Hair colour | | | Eye colour | | |
|:---:|:---:|:---:|:---:|:---:|:---:|:---:|:---:|
| Male | Female | Fair | Brown | Black | Blue | Green | Brown |
| 1 | 0 | 1 | 0 | 0 | 1 | 0 | 0 |
| 1 | 0 | 0 | 0 | 1 | 0 | 0 | 1 |
| 0 | 1 | 1 | 0 | 0 | 0 | 1 | 0 |
| 0 | 1 | 1 | 0 | 0 | 1 | 0 | 0 |
| 1 | 0 | 0 | 1 | 0 | 0 | 0 | 1 |
| 1 | 0 | 0 | 0 | 1 | 0 | 1 | 0 |
| 0 | 1 | 0 | 1 | 0 | 0 | 1 | 0 |
| 0 | 1 | 0 | 1 | 0 | 0 | 0 | 1 |
| 1 | 0 | 1 | 0 | 0 | 1 | 0 | 0 |
| 1 | 0 | 0 | 1 | 0 | 1 | 0 | 0 |

The principal components analysis is carried out as described in Section 6.5.1. If the singular value decomposition of $\mathbf{X}$ is denoted by $\mathbf{X} = \mathbf{T\Theta U'}$, the objects are represented by points with coordinate values given by

$$\mathbf{G} \equiv \mathbf{T\Theta} \qquad (6.27)$$

(compare Equation (6.19)). Now,

$$\mathbf{G} = \mathbf{XU} = p^{-1/2}\mathbf{YC}^{-1/2}\mathbf{U}.$$

Hence,

$$\mathbf{G} = \mathbf{YH}/p \qquad (6.28)$$

if the matrix $\mathbf{H}$, specifying the coordinates of points corresponding to the category levels, is defined by

$$\mathbf{H} \equiv p^{1/2}\mathbf{C}^{-1/2}\mathbf{U}. \qquad (6.29)$$

Equation (6.28) shows that the g-point of an object is located at the centroid of the set of h-points representing the category levels describing the object. This shows how the interactions between objects and category levels are depicted in the joint configuration of points. In addition, relationships within the set of objects and within the set of category levels can also be inferred from the graphical representation. For example, the distance between

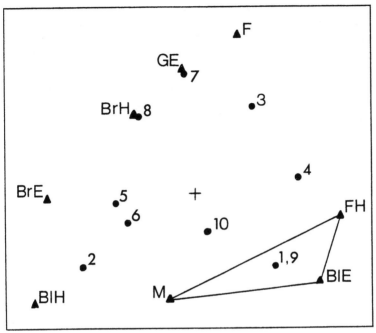

Figure 6.12 *A multiple correspondence analysis of the data presented in Tables 6.3 and 6.4. Codes of category levels: M = male; F = female; FH = fair hair; BrH = brown hair; BlH = black hair; BlE = blue eyes; GE = green eyes; BrE = brown eyes.*

g-points corresponds to a form of chi-squared distance between the rows of the matrix $Y$: if a variable belongs to the $r$th level in the $i$th object and the $s$th level in the $j$th object, the contribution of this variable to the squared distance between the $i$th and $j$th objects is given by $p^{-1}(c_r^{-1} + c_s^{-1})$, where $c_r$ denotes the number of occurrences of the $r$th level of the variable in the data set. A further property of the graphical representation is that each variable has the centroid of its set of category level points, weighted by their relative frequencies, located at the origin of coordinates (Gower and Hand, 1996, Section 4.3).

These properties are illustrated in Fig. 6.12, which displays the first two dimensions of a multiple correspondence analysis of the data presented in Tables 6.3 and 6.4. For example, the category levels 'fair hair' and 'blue eyes' are seen to be positively associated; the individuals 1 and 9 are located at the centroid of the level points 'male', 'fair hair' and 'blue eyes', which describe both of them; and the weighted centroid of the 'blue eyes', 'green eyes' and 'brown eyes' points (weighted in the ratios 4:3:3) is located at the origin, shown by a '+'.

This section has presented only a very brief introduction to selected aspects of biplots. A fuller discussion of these and other types of biplot is presented by Gower and Hand (1996).

# CHAPTER 7

# Cluster validation and description

## 7.1 Introduction

The subject of classification used to be seen as being concerned almost exclusively with the *exploratory* analysis of data sets. There were claims that the main criteria for assessing a classification were its interpretability and usefulness. There are clearly dangers in such an approach: the human brain is quite capable of providing *post hoc* justifications of results of dubious validity. In recent years, there has been an increased awareness of the fact that a classification can impose unwarranted structure on a data set; the partition shown in Fig. 3.5 provides a simple example of inappropriate class structure being imposed on a data set by a clustering algorithm.

This chapter describes ways of assessing the validity of classifications that have been obtained from the application of a clustering procedure or have been otherwise specified, and of carrying out investigations of how they relate to independently-provided information; the presentation draws on that given in Gordon (1998). The assessment has been carried out at several different levels of formality. Some less formal approaches are summarized later in this section, while Section 7.2 describes methodology that involves the testing of hypotheses about the class structure present in data. The level of formality that is appropriate depends on the nature of the data and the investigation, and on the size of the data set; in particular, it seems unlikely that the more elaborate methods of analysis described in this chapter will be relevant for assessing very large data sets. The final section in the chapter considers ways of providing informative descriptions of classes that have been found to be valid.

In addition to offering advice on the selection of appropriate clustering procedures, Section 4.3 mentions two general approaches to combining the output from two or more methods of analysing the same data set. First, one can superimpose a classification structure – such as a partition or a minimum spanning tree – onto a graphical representation, as shown for example in Figs. 3.2 and 3.8. Secondly, one can synthesize the information contained in two or more partitions or hierarchical classifications to obtain a consensus classification, using methodology described in Section 5.6 (for partitions) or Section 4.4 (for hierarchical classifications). In each case, the hope is that the results are less likely to be an artifact of a single method of analysis and more likely to provide a reliable summary of any class structure that is present in the data.

The results of a classification can also be assessed by their stability. This can be investigated by reanalysing a modified version of the data set and noting the extent to which the new classification differs from the original one. The modified data set can be obtained by slightly perturbing the values of the variables or by weighting or deleting some of the objects (e.g. Rand, 1971; Gnanadesikan, Kettenring and Landwehr, 1977; Gordon and De Cata, 1988; Jolliffe, Jones and Morgan, 1988; Cheng and Milligan, 1996).

An alternative approach involves investigating the extent to which similar results are obtained from analyses of subsets of the data (McIntyre and Blashfield, 1980; Morey, Blashfield and Skinner, 1983; Breckenridge, 1989). Such replication studies proceed as follows:

1. The data set is randomly divided into two subsets, $A$ and $B$, say

2. An appropriate clustering procedure is used to partition subset $A$ into a specified number, $c$, of classes

3. Each object in $B$ is assigned to the class in this partition to which it is 'closest', where possible definitions of the closeness of an object to a class include its distance to the class centroid or the smallest dissimilarity between it and some member of the class; this provides a partition of $B$ into no more than $c$ classes

4. The same clustering procedure is used to partition $B$ into $c$ classes

5. The two partitions of $B$ are compared; the greater that the level of agreement is, the more confidence that one may have in the validity of the results.

Further discussion of these methods of assessment is given by Milligan (1996).

## 7.2  Cluster validation

Exploratory analyses of data are commonly conducted in statistics, for example to aid in model formulation. Such analyses are generally followed by confirmatory analyses, which are carried out on *different* data sets. This approach has rarely been followed in classification studies, at least partly because interest often resides solely in the set of objects under investigation and not in some larger collection of objects, from which the data set is regarded as providing a representative sample. Even if this is the case, there are occasions on which a confirmatory stage of analysis could be beneficial. However, because the interest is often in a single data set, which it is considered inappropriate to subdivide for the two stages of the analysis, cluster generation and cluster validation are generally carried out on the same set of objects. This has implications for the manner in which tests are carried out: thus, the classes in a partition provided by a clustering algorithm will tend to be more homogeneous than classes in a randomly-generated partition, and this fact needs to be taken into account when evaluating the null distributions of test statistics.

It is relevant to identify three different types of validation test, based on *external*, *internal* and *relative* indices of clustering tendency (Jain and Dubes, 1988, Chapter 4):

- External tests compare a classification, or a part of a classification, with information that was not used in constructing the classification

- Internal tests compare a (part of a) classification with the original data set

- Relative tests compare several different classifications of the same set of objects

Much of the material in this section is concerned with internal cluster validation tests, but some comments on external and relative validation are also included. The first subsection describes null models of the absence of class structure, and the following four subsections categorize cluster validation tests by the type of class structure that is under study, summarizing tests for

(i) the complete absence of class structure

(ii) the validity of an individual cluster

(iii) the validity of a partition

(iv)  the validity of a hierarchical classification

A final subsection discusses limitations of more formal approaches to internal cluster validation.

### 7.2.1  Null models

Sneath (1967) identifies three classes of null model, based on various kinds of randomness associated with a pattern matrix, a dissimilarity matrix and a tree diagram. A discussion of random tree models is postponed to Section 7.2.5. This section describes five main types of null model; the first three make assertions about a pattern matrix, the fourth postulates a model for a dissimilarity matrix, and the fifth may be used with either a pattern matrix or a dissimilarity matrix.

### Poisson model

This model assumes that the objects can be represented by points that are uniformly distributed in some region $A$ of $p$-dimensional space. It has also been called the uniformity hypothesis (Bock, 1985) and the random position hypothesis (Jain and Dubes, 1988, Chapter 4). $A$ has usually been chosen to be the unit $p$-dimensional hypercube or hypersphere (e.g. Zeng and Dubes, 1985; Hartigan and Mohanty, 1992). An alternative is to specify $A$ to be the convex hull of the points in the data set (Smith and Jain, 1984; Bock, 1985), justifying this choice by reference to the result that if points are randomly generated within a convex region of the plane, an estimate of the boundary of the region is provided by a uniform dilation of the convex hull of the set of points about its centroid (Ripley and Rasson, 1977).

The rationale of such *data-influenced* null models is that they allow the construction of tests which are less influenced by unimportant differences between model and data, such as the region within which the points are located. However, algorithms for finding the convex hull of a set of points in $p$-dimensional space (e.g. Chand and Kapur, 1970; Edelsbrunner, 1987, Chapter 8) and for generating data within such regions (e.g. Dobkin and Lipton, 1976; Rubin, 1984; Chazelle, 1985) make heavy demands on computing resources. Further, an algorithm proposed by Smith and Jain (1984) for generating points within a region approximating the convex hull can perform poorly in 'empty' regions close to the convex hull. Thus, use of the convex hull for the boundary of the Poisson model is limited to small values of $p$, although the development of more efficient algorithms or an increase in computing power may relax this restriction in the future.

*Unimodal model*

Even if there is only a single cluster in the data set, it is possible for objects near the centre of this cluster to be located closer to one another than are objects near the boundary of the cluster. The unimodal model postulates that the joint distribution of the variables describing the objects is unimodal. There are many possible unimodal distributions; a multivariate normal distribution with identity covariance matrix has usually been specified (e.g. Rohlf and Fisher, 1968; Gower and Banfield, 1975; Hartigan and Mohanty, 1992). Data-influenced unimodal models have also been investigated, for example ones that allow the data to specify the covariance matrix of the normal distribution (Gordon, 1996c).

*Random permutation model*

This model considers independently permuting the entries in each of the $p$ columns of the pattern matrix. There are $(n!)^{p-1}$ essentially different matrices, and the model regards each of them as equally likely (e.g. Harper, 1978). This model ignores the correlation between variables, and generates pseudo-objects within a hyperrectangular region of space. Further, clusters in the original data set that are more homogeneous than those found in many pseudo data sets need not necessarily have a high degree of absolute homogeneity.

*Random dissimilarity matrix model*

This model assumes that the elements of the lower triangle of the dissimilarity matrix are ranked in random order, all $(n(n-1)/2)!$ rankings being regarded as equally likely (Ling, 1973a). It has also been called the random graph hypothesis (Jain and Dubes, 1988, Chapter 4): if each object is represented by a vertex in a graph and an edge links the $i$th and $j$th vertices if the dissimilarity $d_{ij}$ is less than some specified threshold value, the edges are inserted into the graph in random order. In variants of the model, a specified number of randomly-selected edges is present in the graph, or each edge has the same probability of being present, independently of the presence or absence of other edges.

Although conceptually attractive, this model has some disadvantages. In that only the ranks of the dissimilarities are taken into account, it would seem to be restricted to use with monotone admissible clustering procedures (see Section 4.3). It also ignores second- and higher-order relationships between the objects: if $d_{ij}$ is small, one might expect $d_{ik}$ and $d_{jk}$ to have similar ranks for most values of $k$. Ling (1973a) argues that clusters that are not significant under this model are unlikely to be deemed significant under any other null model. However, while it might provide a relevant null model when the dissimilarity matrix is provided directly, it does not seem to be appropriate for use with dissimilarities that have been constructed from a pattern matrix.

*Random labels model*

This model assumes that all permutations of the labels of the objects are equally likely (Jain and Dubes, 1988, Chapter 4), and compares the observed value of a test statistic with the distribution of the $n!$ values that it takes under permutation of the labels. It shares some of the disadvantages of the random permutation model described above. Hubert (1987, Chapters 4, 5) presents a detailed discussion of tests based on this model.

### 7.2.2 Tests of the absence of class structure

Many tests of the absence of class structure have been proposed, and this section describes only a selection of them; fuller reviews are given by Bock (1996) and Gordon (1998).

Tests of the Poisson null model have been based on the following test statistics: the number of interpoint distances less than a specified threshold value (Strauss, 1975; Kelly and Ripley, 1976; Saunders and Funk, 1977); the largest nearest neighbour distance within the set of objects (Bock, 1985); and comparisons between ($i$) the distance from a randomly-specified position to the nearest object, and ($ii$) the distance from that object (or a randomly-chosen object) to its nearest neighbour (Cross and Jain, 1982; Panayirci and Dubes, 1983; Zeng and Dubes, 1985).

Tests of the random dissimilarity matrix model have been based on the following test statistics: the number of edges needed before the graph comprises a single component (Rapoport and Fillenbaum, 1972; Schultz and Hubert, 1973; Ling, 1975; Ling and Killough, 1976); the number of components in the graph when it contains a specified number of edges (Ling, 1973b; Ling and Killough, 1976); and the size of clusters when the objects are partitioned into two clusters using the single link criterion (Van Cutsem and Ycart, 1998).

Several tests are based on the minimum spanning tree. The distribution of its edge lengths has been assessed under the Poisson model (Hoffman and Jain, 1983) and under a multivariate normal model (Rohlf, 1975b). Smith and Jain (1984) suggest adding randomly-located points to the data set and using a test proposed by Friedman and Rafsky (1979), based on the number of edges in the minimum spanning tree that link original and added points. Rozál and Hartigan (1994) describe a test based on minimum spanning trees that are constrained to have non-increasing edge lengths on all paths to the root node(s) corresponding to the cluster centre(s).

Other tests are based on searching for 'gaps' or multimodality in the data set. Hartigan and Mohanty (1992) investigate the number of objects in the smaller subclass of each single link class, defining their test statistic to be the largest value that this number takes in the set of all single link classes and simulating its distribution under Poisson and unimodal null models. Müller and Sawitzki (1991) and Polonik (1995) describe tests based on comparing the amounts of probability mass exceeding various threshold values when there are $c$ modes in the distribution; this approach is currently computationally feasible only for small values of the dimensionality, $p$. Hartigan (1988) estimates departures from unimodality along the edges of an optimally-rooted minimum spanning tree.

One might expect an analysis of data to begin with one or more tests of the absence of class structure, and to proceed to obtaining a classification only if the null model(s) were rejected. This has rarely been what has occurred. Reasons why such tests have not been carried out may include that investigators:

- are confident that the data do contain distinct classes of objects

- are interested solely in obtaining a dissection of the data set and/or regard cluster validation tests as irrelevant

- intend subsequently to validate the classification that is obtained, and realize that a two-stage testing procedure would complicate evaluation of the significance level of any test

It should also be noted that some of the tests described above require a classification to be obtained, and can thus be regarded as providing ways of validating particular types of class structure.

### 7.2.3 Assessing individual clusters

One approach to assessing clusters has been to specify properties that they are required to satisfy in order to be regarded as valid; several examples of such 'ideal' clusters are defined in Section 3.4.3. However, it may be difficult to specify a relevant definition of what constitutes an ideal cluster for a particular data set.

More widely applicable methodology involves the definition of an index measuring the adequacy of a cluster, and establishing how likely a given value of the index is under some null model that there is no class structure in the data. Two general approaches within this framework are cluster validity profiles and Monte Carlo validation.

### Cluster validity profiles

Bailey and Dubes (1982) define measures of the isolation and cohesion of a cluster, $C$, as follows. Consider the random graph hypothesis, in which $m$ ($1 \leq m \leq n(n-1)/2$) edges are present in a graph based on $n$ vertices, with each vertex representing a different object in the data set. Let $S_m$ denote the set of edges $(i, j)$ included in the graph, and define raw indices giving the numbers of edges $(i)$ linking members of $C$, and $(ii)$ between members of $C$ and objects outside $C$:

$$W_C(m) = \text{card}\{(i,j)|i, j \in C, (i,j) \in S_m\} \qquad (7.1)$$
$$B_C(m) = \text{card}\{(i,j)|i \in C, j \notin C, (i,j) \in S_m\}. \qquad (7.2)$$

When attention is restricted to considering the edges corresponding to the $m$ smallest dissimilarities in the data set, large values of $W_C(m)$ indicate that cluster $C$ is compact and low values of $B_C(m)$ indicate that it is isolated. For a cluster $C$ of size $r$ that is specified without reference to any clustering procedure, the distributions of these indices under the random graph hypothesis are hypergeometric:

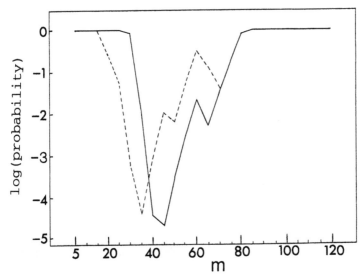

Figure 7.1 *Cluster validity profiles for assessing the compactness of two clusters of the acoustic confusion data:* $\log_{10}(Q_C(m))$ *is plotted against the number of edges, m, for m = 5 (5) 120. The results for cluster $C_1$ are shown by the unbroken line, and the results for cluster $C_2$ are shown by the dashed line.*

$$\text{prob}(W_C(m) = w) = \binom{r_2}{w}\binom{n_2 - r_2}{m - w}/\binom{n_2}{m} \quad (7.3)$$

$$\text{prob}(B_C(m) = b) = \binom{r(n - r)}{b}\binom{n_2 - r(n - r)}{m - b}/\binom{n_2}{m}, \quad (7.4)$$

where $n_2 \equiv n(n - 1)/2$ and $r_2 \equiv r(r - 1)/2$.

If, however, $C$ has been obtained from the results of a clustering procedure, these probabilities need to be modified. By restricting attention to $r$-vertex subsets of the graph for which a raw index takes its optimal value, bounds for the probabilities of the best-case values can be obtained by multiplying the right hand sides of Equations (7.3) and (7.4) by $n!/(r!(n - r)!)$ (Bailey and Dubes, 1982). Thus, for a given value of $m$, indices of the compactness $(Q_C(m))$ and isolation $(I_C(m))$ of a cluster $C$ that has values of the raw indices equal to $w_C$ and $b_C$ are given by bounds on the

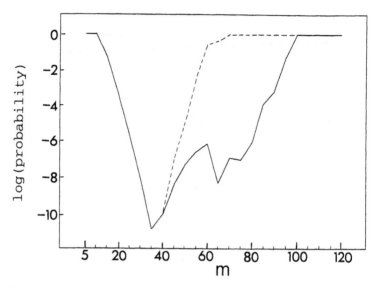

Figure 7.2 *Cluster validity profiles for assessing the isolation of two clusters of the acoustic confusion data:* $\log_{10}(I_C(m))$ *is plotted against the number of edges, $m$, for $m = 5$ (5) 120. The results for cluster $C_1$ are shown by the unbroken line, and the results for cluster $C_2$ are shown by the dashed line.*

probabilities of the observed value or a more extreme value under a random graph hypothesis:

$$Q_C(m) \equiv \min\{ \binom{n}{r} \sum_{w=w_C}^{r_2} \binom{r_2}{w} \binom{n_2 - r_2}{m - w} / \binom{n_2}{m}, 1\} \qquad (7.5)$$

and

$$I_C(m) \equiv \min\{ \binom{n}{r} \sum_{b=0}^{b_C} \binom{r(n - r)}{b} \binom{n_2 - r(n - r)}{m - b} / \binom{n_2}{m}, 1\}. \qquad (7.6)$$

Plots of these indices against $m$ (so-called cluster validity profiles) allow assessment of the compactness and isolation of cluster $C$ at different levels of the classification.

This methodology is illustrated by application to two of the clusters identified in the single link analysis of the acoustic confusion data (see Fig. 4.7), namely

$$C_1 \equiv \{pa, ka, ta, fa, thin, sa, sha\};$$
$$C_2 \equiv \{ba, va, that, za, da, ga, zha\}.$$

In this analysis, some ties in smaller pairwise similarities were arbitrarily broken, but the choices that were made have little effect on the results. The profiles are shown in Figs. 7.1 and 7.2, in which the indices have been evaluated for every fifth value of $m$, and are plotted on a logarithmic scale. It is seen that both clusters are perceived as being compact for values of $m$ between 30 or 35 and 70, and that $C_1$ is perceived as being isolated for a greater range of values of $m$ than $C_2$.

Bailey and Dubes (1982) also consider the application of other types of probability profile to cluster validation.

*Monte Carlo validation*

Monte Carlo tests (Barnard, 1963; Hope, 1968) provide a way of determining whether or not a data set may be regarded as conforming to a null hypothesis. A relevant test statistic is defined, data sets (of the same size as the original one) are simulated under the null hypothesis and classified, and the proportion of them that provide values of the test statistic that are at least as extreme as the observed value is noted.

$U$ statistics (Mann and Whitney, 1947) provide relevant test statistics for assessing the adequacy of a cluster, combining the concepts of its compactness and isolation. Let

$$U_{ijkl} = \begin{cases} 0 & \text{if } d_{ij} < d_{kl} \\ \frac{1}{2} & \text{if } d_{ij} = d_{kl} \\ 1 & \text{if } d_{ij} > d_{kl}. \end{cases} \tag{7.7}$$

where $(d_{ij})$ is a matrix of pairwise dissimilarities within a set of $n$ objects. For a cluster $C$ of size $r$, let

- $W \equiv \{(i, j) | i, j \in C, i < j\}$, the set of $r(r - 1)/2$ within-cluster pairs, and

- $B \equiv \{(k, l) | k \in C, l \notin C\}$, the set of $r(n - r)$ between-cluster pairs.

The global $U$ index, $U_G$, is defined by:

$$U_G \equiv \Sigma_{(i,j)\in W}\Sigma_{(k,l)\in B}U_{ijkl}. \tag{7.8}$$

The local $U$ index, $U_L$, is defined by:

$$U_L \equiv \Sigma_{i\in C}\Sigma_{j\in C\setminus\{i\}}\Sigma_{k\notin C}U_{ijik}. \tag{7.9}$$

The smaller the value of these indices, the better the cluster is regarded as being. Clusters for which the indices take the value zero correspond to two types of 'ideal' cluster defined in Section 3.4.3: thus, an $L^*$-cluster has $U_G = 0$, and a ball cluster has $U_L = 0$.

These $U$ statistics can be used in a Monte Carlo test of the validity of clusters that have been obtained from a classification. The test proceeds as follows (Gordon, 1994):

1. Data sets of size $n$ are randomly generated under a null hypothesis of the absence of class structure, such as the first four models described in Section 7.2.1, and are classified using the same clustering procedure that was used to analyse the original data.

2. The resulting clusters are identified. If more than one cluster containing the same number of objects is found, all but a randomly-selected one are discarded. For each cluster containing $r$ objects that is retained, the value $U(r)$ of the test statistic is stored ($r = 2, 3, ..., n-1$).

3. Steps 1 and 2 are repeated until $(m-1)$ values of $U(r)$ have been obtained for all values of $r$ for which there was a cluster of size $r$ in the classification of the original data.

4. For each value of $r$, the $(m-1)$ values of $U(r)$ are ranked. If a cluster of size $r$ in the original classification has a value of $U$ that is less than the $j$th smallest value of $U(r)$, the null hypothesis can be rejected (and the cluster be deemed valid) at the 100 $(j/m)\%$ level of significance.

This approach has several drawbacks that are common to 'multiple comparisons' tests: thus, tests on clusters in the same classification are not independent of one another, and – because many different tests are being carried out – some clusters are likely to be deemed valid even if the null hypothesis is true. Conversely, the test may fail to detect genuine clusters. This is particularly likely to be a problem for smaller values of the cluster size, $r$: if the $j$th smallest value of $U(r)$ is zero, a cluster of size $r$ cannot be deemed significant at the 100 $(j/m)\%$ level of significance.

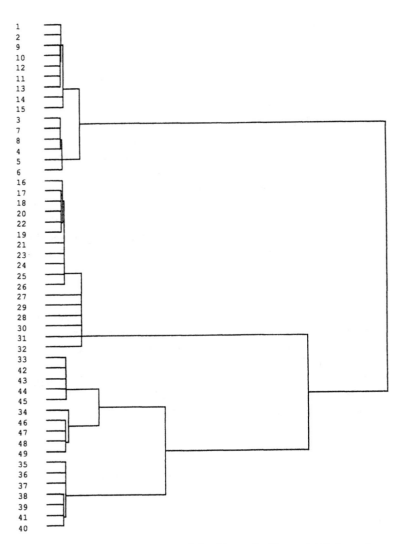

Figure 7.3  *A parsimonious tree of the Abernethy Forest 1974 data, showing only those internal nodes in the incremental sum of squares hierarchical classification that were deemed to be valid in a Monte Carlo
test. Reprinted from A. D. Gordon, Identifying genuine clusters in a
classification, Computational Statistics & Data Analysis, 18, 561–81,
Copyright (1994), with permission from Elsevier Science.*

This methodology is illustrated by an assessment of the validity of the clusters in the hierarchical classification of the Abernethy Forest 1974 data provided by the incremental sum of squares clustering criterion, for which the dendrogram is shown in Fig. 4.9. The statistic $U_G$ was used in the test, and $m$ was chosen to be 100. Two separate investigations were carried out, in which data were generated under a Poisson model within a nine-dimensional hypercube and under a spherical Normal model. It is not possible to assess the validity of clusters of size two or three for the reason given in the preceding paragraph, but all but two of the other clusters are deemed to be valid at the 3% level of significance under both null models. It can be informative to summarize such results by highlighting internal nodes corresponding to significant clusters (Lerman, 1980) or by showing a parsimonious tree (see Section 4.2.4) that retains only the validated clusters; the parsimonious tree for the Abernethy Forest 1974 data is shown in Fig. 7.3.

*External validation of clusters*

Less elaborate tests are available of the validity of a cluster $C$ if it has been specified without reference to any clustering procedure. Thus, the probability distributions under the random graph hypothesis of Bailey and Dubes's (1982) raw indices $W_C(m)$ and $B_C(m)$ are given in Equations (7.3) and (7.4), from which significance levels corresponding to observed values of $w_C$ and $b_C$ can be obtained by summing over values that are at least as extreme as the observed ones:

$$\text{prob}(W_C(m) \geq w_C) = \sum_{w=w_C}^{r_2} \binom{r_2}{w} \binom{n_2 - r_2}{m - w} / \binom{n_2}{m} \quad (7.10)$$

and

$$\text{prob}(B_C(m) \leq b_C) = \sum_{b=0}^{b_C} \binom{r(n-r)}{b} \binom{n_2 - r(n-r)}{m - b} / \binom{n_2}{m}. \quad (7.11)$$

The distribution of other test statistics under various null models can also be evaluated: the analysis proceeds as in the Monte Carlo test described earlier in this section, with the crucial difference that the subsets of size $r$ are randomly selected from the simulated data sets, rather than being obtained from a classification.

The cluster can also be assessed by comparing it with the other subsets of size $r$ in the (genuine) data set, under the random label model that each of the $n!$ permutations of the labels of the $n$ objects is equally likely. For the assessment of individual clusters, this model is equivalent to the null hypothesis that each of the $n!/(r!(n-r)!)$ subsets of size $r$ is equally likely to have been selected. However, even if $C$ turns out to be one of the most homogeneous and/or isolated subsets in the set of all subsets of size $r$, it does not necessarily follow that it has a high absolute degree of homogeneity and/or isolation.

### 7.2.4 Assessing partitions

Several different types of investigation can be relevant when the validity of partitions is being assessed, addressing the following questions:

1. Is there a close correspondence between two independently-derived partitions of the same set of objects?

2. Which of a set of partitions agrees best with an externally-provided partition?

3. Does a specified partition into $c$ (say) clusters comprise compact and isolated clusters?

4. When a clustering procedure provides partitions of data into $c$ clusters for several different values of $c$, which is the most appropriate partition?

5. Does a partition into $c$ clusters obtained from the output of a clustering procedure comprise compact and isolated clusters?

The first three questions involve the comparison of an externally-provided partition with, respectively, a partition, a set of partitions, and a data set. The last two questions involve the assessment of partitions provided by a clustering procedure.

The first question is concerned with external validation. Several different indices have been proposed for the comparison of partitions. Consider two partitions of the same set of $n$ objects: $P_1 \equiv \{C_{1k}(k = 1, ..., c_1)\}$ and $P_2 \equiv \{C_{2k}(k = 1, ..., c_2)\}$. The resemblance between the partitions can be assessed using information contained in the $c_1 \times c_2$ cross-classification table $(n_{ij})$, where

$n_{ij}$ denotes the number of objects belonging to both $C_{1i}$ and $C_{2j}$ $(i = 1, ..., c_1; j = 1, ..., c_2)$.

The $\binom{n}{2}$ distinct pairs of objects can be categorized into three different types:

- those pairs belonging to the same class in $P_1$ and to the same class in $P_2$

- those pairs belonging to a different class in $P_1$ and to a different class in $P_2$

- those pairs belonging to the same class in one of the partitions and to a different class in the other partition

Let $A$ (for 'agreement') denote the total number of pairs of objects of the first or second type above, and $D$ (for 'disagreement') denote the number of pairs of the third type. Defining

$$n_{i.} \equiv \Sigma_{j=1}^{c_2} n_{ij}$$
$$\text{and } n_{.j} \equiv \Sigma_{i=1}^{c_1} n_{ij},$$

it can be shown that

$$A = \binom{n}{2} + 2 \sum_{i=1}^{c_1} \sum_{j=1}^{c_2} \binom{n_{ij}}{2} - [\sum_{i=1}^{c_1} \binom{n_{i.}}{2} + \sum_{j=1}^{c_2} \binom{n_{.j}}{2}]. \quad (7.12)$$

Several functions of $A$ and/or $D$ have been used as measures of the resemblance or difference between two partitions: for example, Rand's (1971) statistic is defined by $R \equiv 2A/(n(n-1))$. A critical assessment of this and other statistics for comparing partitions is given by Hubert and Arabie (1985), who argue that Rand's statistic should be corrected for chance, so as to ensure that its maximum value is 1 and its expected value is zero when the partitions are selected at random (subject to the constraint that the row and column totals of $(n_{ij})$ are fixed). The corrected Rand statistic is

$$R_{HA} \equiv \frac{\Sigma_{i=1}^{c_1} \Sigma_{j=1}^{c_2} \binom{n_{ij}}{2} - \Sigma_{i=1}^{c_1} \binom{n_{i.}}{2} \Sigma_{j=1}^{c_2} \binom{n_{.j}}{2} / \binom{n}{2}}{[\Sigma_{i=1}^{c_1} \binom{n_{i.}}{2} + \Sigma_{j=1}^{c_2} \binom{n_{.j}}{2}]/2 - \Sigma_{i=1}^{c_1} \binom{n_{i.}}{2} \Sigma_{j=1}^{c_2} \binom{n_{.j}}{2} / \binom{n}{2}}.$$
$$(7.13)$$

The resemblance between two independent partitions of the same set of objects can be assessed by comparing their value of $R_{HA}$ with its distribution under the random label hypothesis. For larger values of $n$, the complete set of $n!$ values of $R_{HA}$ would not be evaluated; instead, the comparison would be made with the values resulting from a randomly-selected subset of permutations.

The second question combines relative and external cluster validation. It can be of interest to compare an externally-specified partition with several different partitions of a set of objects in order to identify the one that provides the closest match. A comparative study of five indices conducted by Milligan and Cooper (1986) concluded that the modified Rand index, $R_{HA}$, was best suited for this kind of investigation.

The third question concerns comparing data with a partition that was obtained independently of them. Several indices measuring the adequacy of a partition were described in Section 3.1. Alternatively, one may use modifications of the $U$ statistics defined in Section 7.2.3; for example, the $U_G$ statistic (Equation (7.8)) may be used, with the difference that now $W$ denotes the set of all within-cluster pairs and $B$ denotes the set of all between-cluster pairs (Lerman, 1970, Chapter 2; 1980). The observed value of the criterion of partition adequacy can then be compared with its values under one of the null models of the absence of class structure described in Section 7.2.1: for the random labels and random permutation models, this involves evaluating the criterion for permuted versions of the original data set; for the other models, data sets are simulated under the null model, and the criterion is evaluated for a partition that is randomly selected from each simulated data set, subject to the constraint that it have the same number and sizes of classes.

The fourth question has already been addressed in Section 3.5, in which investigations of the efficacy of several stopping rules were described.

The fifth and final question addresses a problem of internal cluster validation: how can one assess the validity of a partition provided by a clustering criterion? As before, a measure of the adequacy of a partition is defined, and its distribution under a null model of the absence of class structure is sought. Some asymptotic theoretical results have been obtained for some partitioning criteria, as reviewed in Bock (1996), but investigations of their relevance for the analysis of finite data sets generally require simulation studies. Monte Carlo tests of the validity of a partition provided by a clustering procedure have been carried out along the lines of the tests for individual clusters that are described in Section 7.2.3

(Arnold, 1979; Milligan and Mahajan, 1980): thus, given a relevant measure of the heterogeneity of a partition into $c$ clusters ($H(c)$, say), the test proceeds as follows:

1. Data sets of size $n$ are generated under a model of the absence of class structure, such as the first four models described in Section 7.2.1, and are partitioned into $c$ clusters using the same clustering procedure that was used on the original data set.

2. The value of $H(c)$ for this partition is noted.

3. Steps 1 and 2 are repeated $(m - 1)$ times.

4. The $(m - 1)$ values of $H(c)$ are ranked. If the original partition has a value of $H(c)$ that is less than the $j$th smallest simulated value of $H(c)$, the null hypothesis can be rejected (and the partition be deemed valid) at the $100\ (j/m)\%$ level of significance.

### 7.2.5 Assessing hierarchical classifications

The following questions can be relevant when the validity of hierarchical classifications is being assessed:

1. Is there a close correspondence between two independently-derived hierarchical classifications of the same set of objects?

2. Does a specified hierarchical classification provide an accurate summary of the relationships within a set of objects?

3. Does a hierarchical classification obtained from the application of a clustering procedure to a set of objects provide an accurate summary of the data?

The first two questions involve the comparison of an externally-specified hierarchical classification with, respectively, a hierarchical classification and a data set. Table 4.5 summarizes several measures of the agreement and discordance between a set of dissimilarities, $(d_{ij})$, and a set of ultrametric distances specifying a hierarchical classification of the same data, $(h_{ij})$; and these can be used to provide indices that are relevant for addressing the second question. By replacing $(d_{ij})$ by another set of ultrametric distances, $(\tilde{h}_{ij})$, one obtains indices that compare two sets of ultrametric distances, which can be used to address the first question. Other relevant indices are reviewed by Rohlf (1982b).

The distributions of such indices under an appropriate null model allow a test of null hypotheses that there is no relationship between two hierarchical classifications or between a hierarchical classification and a data set. One such null model is the random label model, but several random tree models, described below, may be more relevant.

The random tree models assume that all possible rooted trees based on $n$ objects are equally likely. There are several different models, depending on whether or not the trees are binary $(B)$, labelled $(L)$, or ranked $(R)$. (Recalling definitions given in Section 4.1: a tree is binary if all of its internal nodes have degree three, except for the root, which has degree two; a non-ranked tree is an $n$-tree, in which only subset information is available; and a ranked tree specifies the global ordering of the heights of the internal nodes.) The labelling of a tree refers only to the labelling of its terminal nodes; in this account, its internal nodes are always considered to be unlabelled.

The number of different trees increases rapidly with $n$ for each type of tree; Murtagh (1984) summarizes recurrence relations for evaluating these numbers, listing them for small values of $n$. Because of the large numbers of possible trees, it is usual to select a random subset of trees from the relevant population. Furnas (1984) critically reviews algorithms that have been proposed for the uniform generation of these (and other) types of tree: general strategies include recursive construction, transformation from a different type of randomly-generated structure, and (given a suitable numbering scheme for a population of trees) selecting the tree corresponding to a uniformly-distributed integer. Using such strategies, algorithms have been proposed for the random generation of, amongst others, $BLR$ (Rohlf, 1983), $BL\bar{R}$ (Rohlf, 1983; Furnas, 1984; Quiroz, 1989), and $\bar{B}L\bar{R}$ (Oden and Shao, 1984) trees. Lapointe and Legendre (1990, 1991) note that if the $(n-1)$ fusion levels in a valued tree appear in the subdiagonal of an $(n \times n)$ matrix, the complete set of all $n(n-1)/2$ ultrametric distances can be obtained by making use of the ultrametric inequality (Equation (4.1)), and prove that a random valued tree with the same fusion levels can be obtained from random permutations of both the set of elements in the subdiagonal and the labels of the objects. Further discussion of the random generation of trees is given by Furnas (1984) and Gordon (1996, Section 8).

Using this methodology, the first two questions can be addressed by comparing the observed value of a relevant index, such as $\gamma$, with its distribution under a random tree hypothesis. The first question requires pairs of random trees to be generated and compared; an asymmetric variant of this question that seeks to investigate the closeness of a tree to a designated tree, would involve the random generation of a single set of trees. Further discussion and examples of such tests are presented by Lapointe (1998), who describes a stepwise procedure that includes both external and internal validation of hierarchical classifications.

However, the assumption that trees should be *uniformly* generated from the relevant population of trees can be questioned if the hierarchical classification has been obtained by application of a clustering procedure, rather than by being specified directly from theoretical considerations. This is because of the tendency of different clustering criteria to provide different shapes of tree. For example, the single link criterion is more likely than the complete link criterion to yield unbalanced trees, and it can be argued that the validity of a hierarchical classification provided by the single link criterion should be assessed under a null model of random single link trees. The distribution of single link trees under the random dissimilarity matrix model is studied by Frank and Svensson (1981), but criticisms of this model have been made earlier in this section. There remains a need for further work in this area.

The third question addresses the internal validation of a hierarchical classification obtained from a clustering procedure. Simulations have been carried out to obtain the distributions (under Poisson, unimodal and random dissimilarity matrix models) of several measures of the distortion imposed on data by a hierarchical classification (Rohlf and Fisher, 1968; Hubert, 1974b; Gower and Banfield, 1975). A related question is considered by Lerman (1970, Chapter 4; 1981, Chapter 3), who investigates the extent to which a dissimilarity matrix satisfies the ultrametric condition that the two largest pairwise dissimilarities in each triple $\{d_{ij}, d_{ik}, d_{jk}\}$ should be equal. The number of dissimilarity values lying strictly between this pair, summed over all triples and standardized by dividing by $\binom{n}{3} \times \binom{n}{2}$, provides a measure of the 'lack of ultrametricity' of a data set. Lerman investigates its properties under sampling from the set of binary pattern matrices with specified row sums.

However, even if a null model of the absence of class structure is rejected, it does not follow that the entire hierarchical classification is validated; as argued in Section 4.2.4, it is often relevant to retain only part of a hierarchical classification. It therefore seems preferable to undertake investigations of the constituent parts of a hierarchical classification, carrying out tests of the validity of partitions or individual clusters.

### 7.2.6 Limitations

This section contains a discussion of limitations of the use of more formal internal validation tests. In a hypothesis testing context, one would like to be able to specify

- relevant null hypotheses of the absence of class structure in a data set
- alternative hypotheses describing departures from such null hypotheses which it is important to detect
- test statistics with known properties, which are effective in detecting class structure

Several disadvantages of the null models described in this chapter have already been mentioned: the random dissimilarity matrix and random permutation models seem inappropriate for the internal validation of classifications of either pattern matrices or dissimilarity matrices derived from them. Gordon (1994, 1996c) investigates the extent to which critical values of $U$ statistics depend on the precise null model that is postulated, noting differences between standard and data-influenced null models. Data-influenced Poisson and unimodal models seem particularly relevant, but it can be difficult to specify them and to generate data under such models. Further, because these models aim to eliminate unimportant differences between model and data, they can be expected to be less likely than standard null models to reject hypotheses of the absence of class structure in a data set.

Several alternative models, specifying the presence of class structure in data, are available. For example, a multimodal distribution for the variables can be specified by postulating that the data have been obtained from a mixture of distributions that differ only in their location parameters (Bock, 1985, 1996; Hartigan and Mohanty, 1992) or from models arising in the analysis of spatial point patterns, such as the Neyman-Scott cluster process (Diggle, 1983, Chapter 4). Investigations have rarely been conducted of the power of tests under such alternative hypotheses (cf. Hartigan and Mohanty, 1992; Van Cutsem and Ycart, 1998), and further work is needed, in order to allow assessment of the general usefulness of various test statistics.

A particularly telling criticism of internal cluster validation tests is that it is only (parts of) classifications that have been obtained from the results of a cluster analysis that are assessed, and it is thus necessary to assume that the data have been classified using an appropriate method of analysis. Given the uncertainties associated with the many decisions that need to be taken before an appropriate clustering procedure is selected (see Section 4.3) and the many different kinds of cluster structure that could be present in data, it could be argued that a formal hypothesis testing approach is too elaborate. The author shares these reservations, but believes that in some instances cluster validation provides a useful component of a classification study.

## 7.3 Cluster description

The final stage in a classification is the description of valid clusters, in order to allow the efficient summarization of data and the assignment of new objects to these clusters.

If the only recorded information about the original objects is a matrix of pairwise dissimilarities $(d_{ij})$, the properties of a cluster can only be described by identifying a few of its objects and/or by using summary measures based on the dissimilarities. It is thus useful to investigate the strength with which an object belongs to the cluster in which it is located, in order to identify 'core' and 'outlying' members of the cluster. Such information can be provided by an examination of configurations of points provided by the graphical procedures described in Chapter 6, by fuzzy clustering (Section 5.1), or by methodology described in the following paragraphs.

Silhouette plots (Rousseeuw, 1987) provide a graphical method of assessing the relative compactness and isolation of clusters, and distinguishing between their core and outlying members. If the $i$th object belongs to cluster $C_r$, which contains $n_r (\geq 2)$ objects, define

$$a(i) = \Sigma_{j \in \{C_r \setminus i\}} d_{ij}/(n_r - 1) \qquad (7.14)$$

$$b(i) = \min_{s \neq r} \{\Sigma_{j \in C_s} d_{ij}/n_s\} \qquad (7.15)$$

$$\text{and } s(i) = \frac{b(i) - a(i)}{\max\{a(i), b(i)\}}. \qquad (7.16)$$

(A similar definition can be given if the relationships between the

Table 7.1 *A silhouette plot for the partition of the acoustic confusion data into three clusters.*

| Object | $s(i)$ | 0000111222333344445555666667778889999<br>0369258147036925814703692581470369 |
|:---:|:---:|:---|
| pa | 0.870 | ***************************** |
| ka | 0.864 | **************************** |
| ta | 0.830 | *************************** |
| fa | 0.798 | ************************** |
| thin | 0.770 | ************************* |
| sa | 0.669 | ********************** |
| sha | 0.652 | ********************* |
| | | |
| va | 0.797 | ************************** |
| that | 0.791 | ************************** |
| ga | 0.781 | ************************** |
| za | 0.777 | ************************* |
| da | 0.774 | ************************* |
| zha | 0.686 | ********************** |
| ba | 0.670 | ********************* |
| | | |
| na | 0.828 | *************************** |
| ma | 0.816 | *************************** |

objects are summarized by a similarity matrix.) If the $i$th object belongs to a singleton cluster, $s(i) \equiv 0$.

The values of $s(i)$ lie between $-1$ and $+1$, with negative values indicating that the object is more similar to members of another cluster, and values near $+1$ indicating that the object strongly belongs to the cluster in which it has been placed. When plotted, the set of values $\{s(i)|i \in C_r\}$ provides the silhouette of cluster $C_r$, and the strength with which each object belongs to $C_r$ is shown in the cluster silhouette. As an illustration, Table 7.1 presents a plot of the silhouettes in the partition of the acoustic confusion data into the following three clusters: $\{pa, ka, ta, fa, thin, sa, sha\}$, $\{va, that, ga, za, da, zha, ba\}$, $\{na, ma\}$. Within each cluster, the objects have been ordered by their values of $s(i)$. For these data, each cluster is indicated as being reasonably compact, none of the objects appearing to be extreme outliers.

Each of the clusters can be summarized by its $k$ objects with largest values of $s(i)$. The case $k = 1$ corresponds to specifying a single representative object or 'prototype' for each cluster. An alternative prototype object is the medoid or star centre of the cluster (see Section 3.3), i.e. the object $m \in C_r$ which provides the minimum value of $\Sigma_{j \in C_r} d_{ij}$. The medoid need not be uniquely defined, and will not be when the cluster contains only two objects.

The properties of clusters can also be studied by evaluating the dissimilarity $d_{im(r)}$ between the $i$th object and the $r$th cluster's prototype, $m(r)$ $(i = 1, ..., n; r = 1, ..., c)$, and plotting these dissimilarities against the objects, labelling the plotting position by the cluster code. In the plot, each object has a different coordinate value, with objects within the same cluster being given similar coordinate values (Gnanadesikan, Kettenring and Landwehr, 1977). For large data sets, the size of the plot can be reduced by allowing some overwriting within clusters (Fowlkes, Gnanadesikan and Kettenring, 1988). This latter type of plot is illustrated for the Abernethy Forest 1974 data. The medoids of a partition of these data into five clusters were obtained in Section 3.3, and are shown circled in Fig. 3.2. The partition of the data into five clusters obtained by minimizing the sum of squares criterion $P(H1, \Sigma)$ was found to consist of the following clusters:

1: $\{1-15\}$; 2: $\{16-32\}$; 3: $\{35-41\}$;
4: $\{33, 42-45\}$; 5: $\{34, 46-49\}$.

Fig. 7.4 shows an investigation of these five clusters. The vertical coordinate values of objects within the same cluster have been specified by adding a small random perturbation to the value of the cluster medoid. The plotting symbol denotes the medoid to which the object-medoid dissimilarity refers. Fig. 7.4 clearly displays the variability within each of the clusters, for example confirming that the third cluster is particularly compact and isolated. In addition, the second cluster appears to contain an outlier, one of its objects being closer to the medoid of the fourth cluster than to its own medoid (object 32: see Section 3.3).

In addition to summarizing properties of clusters, plots such as those shown in Table 7.1 and Fig. 7.4 have also been used to assist the determination of the number of clusters in the data and – when the dissimilarities have been constructed from a pattern matrix – to investigate the influence of selected variables.

Several measures of the variability of a cluster are given by het-

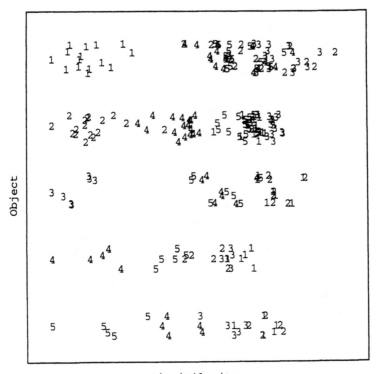

Dissimilarity

Figure 7.4 *A plot showing the dissimilarity between each object and the medoids of each of the clusters in a partition of the Abernethy Forest 1974 data into five clusters. Objects in the same cluster have similar coordinate values.*

erogeneity criteria $H3 - H5$ in Section 3.1. However, for the purpose of assigning new objects, it can be more informative to retain some information about the distribution of the values of the dissimilarities between each object and the cluster prototype. A new object can then be compared with each of the prototypes, and either be assigned to the cluster whose prototype is most similar to it, or be deemed to be a representative of a new cluster if it is sufficiently different from all of the prototypes.

These methods of analysis can also be used when the information about the set of objects is provided by a pattern matrix, from which a dissimilarity matrix is derived. In this case, however, the prototype of a cluster need not be restricted to being one of the original objects, but can be defined by a summary of the set of values taken by each variable in the cluster: for example, the mean of continuous variables, the median of ordinal variables, and the mode of nominal variables. Further, the assignment of new objects can be made to depend not only on their dissimilarities with each of the set of prototypes, but also on the within-cluster variability: one implementation of this idea is presented in Section 3.4.2.

However, when a pattern matrix is available, there is a wider range of options. The partition of the objects into classes (possibly after the deletion of some outlying objects) is regarded as given, and one seeks to derive 'rules' that allow the assignment of new objects to the class which they most resemble. Many different methods have been proposed for constructing such rules, not all of which provide an explicit description of each class; however, the set of methods known collectively as decision trees (Breiman et al., 1984; Feng and Michie, 1994; Ripley, 1996, Chapter 7) do provide such class descriptions. As outlined in Section 5.4.2, at each internal node of a decision tree, a question is asked about one of the variables describing the set of objects. The answer to the question specifies which of the branches is followed and hence which offspring node is next encountered. Each terminal node is associated with one of the classes; the results of the tests encountered before reaching it provide a definition of the class (or possibly a part of the definition, if more than one terminal node is associated with the class).

In the construction of decision trees, the variable that is selected at a given node, and the way in which its values are categorized, are chosen so as to ensure that the resulting subsets are as 'pure' as possible. If $p_r$ denotes the proportion of objects reaching a node that belong to the $r$th class ($r = 1, ..., c$), two commonly used measures of node impurity are the Gini index,

$$I_G(\mathbf{p}) \equiv 1 - \Sigma_{r=1}^c p_r^2,$$

and the entropy,

$$I_E(\mathbf{p}) \equiv \Sigma_{r=1}^c p_r \mathrm{log} p_r.$$

The division is chosen so as to maximize the reduction in the measure of impurity. Attention is often restricted to binary divisions of variables: for example, the values of the $i$th variable can be categorized as $(v_i \leq v_i^0)$ and $(v_i > v_i^0)$ if it is quantitative, or as

$(v_i \in \{c_{\pi(1)}, ..., c_{\pi(t)}\})$ and $(v_i \in \{c_{\pi(t+1)}, ..., c_{\pi(s)}\})$ for an $s$-state nominal variable.

It is rarely the case that each of the terminal nodes of a decision tree can be reached only by objects belonging to a single class of the partition used in the construction of the tree. Even if this could be achieved by having a sufficiently large number of internal nodes in the tree, the final divisions are regarded as less reliable for assignment purposes, and it is common not to include them. This illustrates the fact that the aims of cluster description and assignment can differ. The aim of cluster description could be:

- solely to allow the assignment of new objects to existing clusters
- to provide a full description of each cluster
- to describe all characteristics of a cluster that distinguish it from other clusters

Thus, decision trees provide succinct descriptions of each cluster in terms of the values of a small number of variables. This suffices for the assignment of new objects, but might not provide a very detailed description of the clusters. By contrast, Section 5.5 explains how one can define a second order symbolic object to represent the set of objects belonging to each cluster. This facility is particularly valuable if one wishes to make use of summaries of individual clusters, without reference to any partition of which they were a part. However, when such clusters are used in conjunction with other clusters (whether or not these were obtained in the same classification), there can be substantial overlap between the definitions of different clusters. It is thus relevant also to consider refining cluster definitions (possibly deleting outlying members of clusters), so as to reduce and preferably eliminate the overlap between the (modified) clusters.

Many proposals for addressing this problem have been made in the fields of machine learning (e.g. Michalski, Carbonnel and Mitchell, 1983), knowledge discovery and data mining (e.g. Fayyad et al., 1996) and conceptual clustering (see Section 5.4). The following presentation draws on papers by Ho, Diday and Gettler-Summa (1988) and Stéphan (1996), which present methodology that can also be used to describe clusters whose constituent objects are (first order) symbolic objects.

As an illustration, Fig. 7.5 portrays a set of objects described by two quantitative variables, with members of a cluster $C_r$, for which a description is sought, being labelled '1', and all objects

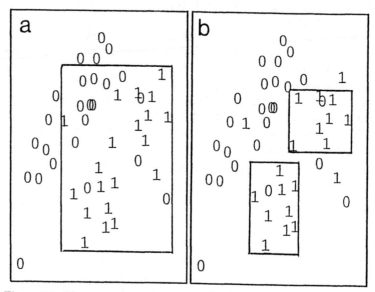

Figure 7.5 *Illustrating the description of a cluster whose constituent objects are labelled '1'. The rule is satisfied by all objects contained in the rectangle(s): (a) all objects labelled '1', but also some objects labelled '0', are included in the description; (b) there is a reduction in the number of objects that are included in the description, both those labelled '0' and those labelled '1'.*

not belonging to $C_r$ being labelled '0'. The second order symbolic object defining $C_r$ is shown by the rectangle in Fig. 7.5(a), which includes all of the members of $C_r$ but also 13 objects not belonging to $C_r$. Fig. 7.5(b) illustrates how the number of 0's that satisfy the definition can be reduced, at the cost of failing to include some of the 1's. For a given subset $S$ of the set $\Omega$ of all objects under consideration, let $R(S)$ denote the subset of objects that are covered by rectangle(s) (more generally, 'rule') $R$. Three relevant criteria for assessing rules for describing clusters are:

(i) the number of 'false positives', i.e. objects belonging to $R(\Omega)$ but not to $C_r$

(ii) the number of 'false negatives', i.e. objects belonging to $C_r$ but not to $R(C_r)$

(iii) the 'volume' of attribute space covered by the rule.

Examples of indices that can be used for the first two criteria are:

$$P_r(R) \equiv \text{card}(R(\Omega \backslash C_r))/\text{card}(R(C_r)) \qquad (7.17)$$

$$N_r(R) \equiv \text{card}(C_r \backslash R(C_r))/\text{card}(C_r). \qquad (7.18)$$

The 'volume' of attribute space can be defined by

$$V \equiv \prod_{k=1}^{p} w_k l_k, \qquad (7.19)$$

where $l_k$ is defined to be $(max_i(x_{ik}) - min_i(x_{ik}))$ for a quantitative variable or the number of states of a categorical variable, and $w_k$ is a weighting factor for the $k$th variable; alternatively, all quantitative variables can be discretized.

Good cluster descriptions have relatively small values of some or all of the three indices (7.17) – (7.19), subject often to the proviso that the description should not be too complicated (i.e. it should not comprise a large number of sub-rules). If the aim is to describe a single cluster, with no attention being paid to its relationship with other clusters, the value of $P_r(R)$ is less relevant. Algorithms for obtaining such cluster descriptions are described by Ho, Diday and Gettler-Summa (1988), Tong and Ho (1991) and Stéphan (1996), and examples of their application are presented in these papers and in Gettler Summa, Périnel and Ferraris (1994).

A supplementary exercise in cluster description involves the investigation of the clusters in order to establish whether or not they can be given substantive interpretations: for example, Arabie and Carroll (1980) describe classes of the acoustic confusion data in terms of whether their constituent consonants were voiced or unvoiced, and were fricatives or stops (see Table 5.2). Such substantive descriptions do not make direct use of data, but require investigators to reflect on the results of classification studies.

# References

Amir, A. and Keselman, D. (1997) Maximum agreement subtree in a set of evolutionary trees: metrics and efficient algorithms. *SIAM Journal on Computing*, **26**, 1656–69.

Anderberg, M. R. (1973) *Cluster Analysis for Applications*, Academic Press, New York.

Anderson, A. J. B. (1966) *A Review of Some Recent Developments in Numerical Taxonomy*, M.Sc. thesis, University of Aberdeen.

Anderson, E. (1960) A semigraphical method for the analysis of complex problems. *Technometrics*, **2**, 387–91.

Arabie, P. (1991) Was Euclid an unnecessarily sophisticated psychologist? *Psychometrika*, **56**, 567–87.

Arabie, P. and Carroll, J. D. (1980) MAPCLUS: a mathematical programming approach to fitting the ADCLUS model. *Psychometrika*, **45**, 211–35.

Arabie, P. and Carroll, J. D. (1989) Conceptions of overlap in social structure, in *Research Methods in Social Network Analysis* (eds L. C. Freeman, D. R. White and A. K. Romney), George Mason University Press, Fairfax, VA, pp. 367–92.

Arabie, P., Carroll, J. D. and DeSarbo, W. S. (1987) *Three-Way Scaling and Clustering*, Sage Publications, Newbury Park.

Arnold, S. J. (1979) A test for clusters. *Journal of Marketing Research*, **16**, 545–51.

Art, D., Gnanadesikan, R. and Kettenring, J. R. (1982) Data-based metrics for cluster analysis. *Utilitas Mathematica*, **21A**, 75–99.

Asimov, D. (1985) The grand tour: a tool for viewing multidimensional data. *SIAM Journal on Scientific and Statistical Computing*, **6**, 128–43.

Aurenhammer, F. (1991) Voronoi diagrams – a survey of a fundamental geometric data structure. *ACM Computing Surveys*, **23**, 345–405.

Bailey, T. A. and Dubes, R. (1982) Cluster validity profiles. *Pattern Recognition*, **15**, 61–83.

Ball, G. H. and Hall, D. J. (1967) A clustering technique for summarizing multivariate data. *Behavioral Science*, **12**, 153–5.

Bandelt, H. J. and Barthélemy, J. P. (1984) Medians in median graphs. *Discrete Applied Mathematics*, **8**, 131–42.

Banfield, C. F. and Bassill, L. C. (1977) Algorithm AS 113. A transfer algorithm for non-hierarchical classification. *Applied Statistics*, **26**, 206–10.

Banfield, J. D. and Raftery, A. E. (1993) Model-based Gaussian and non-Gaussian clustering. *Biometrics*, **49**, 803–21.

Barnard, G. A. (1963) Discussion of a paper by M. S. Bartlett. *Journal of the Royal Statistical Society*, **B 25**, 294.

Barnett, J. A., Bascomb, S. and Gower, J. C. (1975) A maximal predictive classification of Klebsielleae and of the yeasts. *Journal of General Microbiology*, **86**, 93–102.

Barrett, M., Donoghue, M. J. and Sober, E. (1991) Against consensus. *Systematic Zoology*, **40**, 486–93.

Barthélemy, J. P. (1988) Thresholded consensus for *n*-trees. *Journal of Classification*, **5**, 229–36.

Barthélemy, J. P. and McMorris, F. R. (1986) The median procedure for *n*-trees. *Journal of Classification*, **3**, 329–34.

Batagelj, V. (1981) Note on ultrametric hierarchical clustering algorithms. *Psychometrika*, **46**, 351–2.

Baulieu, F. B. (1989) A classification of presence/absence based dissimilarity coefficients. *Journal of Classification*, **6**, 233–46.

Baulieu, F. B. (1997) Two variant axiom systems for presence/absence based dissimilarity coefficients. *Journal of Classification*, **14**, 159–70.

Beale, E. M. L. (1969) Euclidean cluster analysis. *Bulletin of the International Statistical Institute*, **43** (2), 92–4.

Becker, R. A. and Cleveland, W. S. (1987) Brushing scatterplots. *Technometrics*, **29**, 127–42.

Becker, R. A., Cleveland, W. S. and Weil, G. (1988) The use of brushing and rotation for data analysis, in *Dynamic Graphics for Statistics* (eds W. S. Cleveland and M. E. McGill), Wadsworth & Brooks/Cole, Belmont, CA, pp. 247–75.

Benzécri, J.-P. (1982) Construction d'une classification ascendante hiérarchique par la recherche en chaîne des voisins réciproques. *Les Cahiers de l'Analyse des Données*, **7**, 209–18.

Berry, B. J. L. (1968) A synthesis of formal and functional regions using a general field theory of spatial behaviour, in *Spatial Analysis: A Reader in Statistical Geography* (eds B. J. L. Berry and D. F. Marble), Prentice-Hall, Englewood Cliffs, NJ, pp. 419–28.

Bertin, J. (1983) *Semiology of Graphics: Diagrams, Networks, Maps* (translated by W. J. Berg), University of Wisconsin Press, Madison, WI.

Bertrand, P. (1995) Structural properties of pyramidal clustering, in *Partitioning Data Sets* (eds I. J. Cox, P. Hansen and B. Julesz), DIMACS Series in Discrete Mathematics and Theoretical Computer Science, Volume 19, American Mathematical Society, Providence, RI, pp. 35–53.

Bezdek, J. C. (1974) Numerical taxonomy with fuzzy sets. *Journal of Mathematical Biology*, 1, 57–71.

Bezdek, J. C. (1981) *Pattern Recognition With Fuzzy Objective Functions*, Plenum Press, New York.

Bezdek, J. C. (1987) Some non-standard clustering algorithms, in *Developments in Numerical Ecology* (eds P. Legendre and L. Legendre), Springer, Berlin, pp. 225–87.

Bhattacharyya, A. (1943) On a measure of divergence between two statistical populations defined by their probability distributions. *Bulletin of the Calcutta Mathematical Society*, 35, 99–109.

Binder, D. A. (1978) Bayesian cluster analysis. *Biometrika*, 65, 31–8.

Birks, H. H. (1970) Studies in the vegetational history of Scotland I. A pollen diagram from Abernethy Forest, Inverness-shire. *Journal of Ecology*, 58, 827–46.

Birks, H. H. and Mathewes, R. W. (1978) Studies in the vegetational history of Scotland V. Late Devensian and early Flandrian pollen and macrofossil stratigraphy at Abernethy Forest, Inverness-shire. *New Phytologist*, 80, 455–84.

Birks, H. J. B. (1976) The distribution of European pteridophytes: a numerical analysis. *New Phytologist*, 77, 257–87.

Bishop, C. M. (1995) *Neural Networks for Pattern Recognition*, Clarendon Press, Oxford.

Bobisud, H. M. and Bobisud, L. E. (1972) A metric for classifications. *Taxon*, 21, 607–13.

Bobrowski, L. and Bezdek, J. C. (1991) $c$-means clustering with the $l_1$ and $l_\infty$ norms. *IEEE Transactions on Systems, Man, and Cybernetics*, 21, 545–54.

Bock, H. H. (1985) On some significance tests in cluster analysis. *Journal of Classification*, 2, 77–108.

Bock, H.-H. (1996) Probability models and hypotheses testing in partitioning cluster analysis, in *Clustering and Classification* (eds P. Arabie, L. J. Hubert and G. De Soete), World Scientific, Singapore, pp. 377–453.

Bock, H.-H. (1998) Clustering and neural networks, in *Advances in Data Science and Classification* (eds A. Rizzi, M. Vichi and H.-H. Bock), Springer, Berlin, pp. 265–77.

Bock, H.-H. et al. (Eds) (1999) *Statistical Analysis of Symbolic Data*, to be published by Springer.

Boorman, S. A. and Olivier, D. C. (1973) Metrics on spaces of finite trees. *Journal of Mathematical Psychology*, 10, 26–59.

Breckenridge, J. N. (1989) Replicating cluster analysis: method, consistency, and validity. *Multivariate Behavioral Research*, 24, 147–61.

Breiman, L., Friedman, J. H., Olshen, R. A. and Stone, C. J. (1984)

*Classification and Regression Trees*, Wadsworth, Belmont, CA.

Bremer, K. (1990) Combinable component consensus. *Cladistics*, **6**, 369–72.

Brossier, G. (1982) Classification hiérarchique à partir de matrices carrées non symétriques. *Statistiques et Analyse des Données*, **7**, 22–40.

Brossier, G. (1990) Piecewise hierarchical clustering. *Journal of Classification*, **7**, 197–216.

Brucker, P. (1978) On the complexity of clustering problems, in *Optimization and Operations Research* (eds R. Henn, B. Korte and W. Oettli), Springer, Berlin, pp. 45–54.

Bruynooghe, M. (1978) Classification ascendante hiérarchique des grands ensembles de données: un algorithme rapide fondé sur la construction des voisinages réductibles. *Les Cahiers de l'Analyse des Données*, **3**, 7–33.

Bryant, J. (1979) On the clustering of multidimensional pictorial data. *Pattern Recognition*, **11**, 115–25.

Bryant, P. G. (1991) Large-sample results for optimization-based clustering methods. *Journal of Classification*, **8**, 31–44.

Bryant, P. and Williamson, J. A. (1978) Asymptotic behaviour of classification maximum likelihood estimates. *Biometrika*, **65**, 273–81.

Buck, S. F. (1960) A method of estimation of missing values in multivariate data suitable for use with an electronic computer. *Journal of the Royal Statistical Society*, **B 22**, 302–6.

Cailliez, F. (1983) The analytical solution to the additive constant problem. *Psychometrika*, **48**, 305–8.

Cailliez, F. and Kuntz, P. (1996) A contribution to the study of metric and Euclidean structures of dissimilarities. *Psychometrika* , **61**, 241–53.

Cain, A. J. (1962) The evolution of taxonomic principles, in *Microbial Classification* (eds G. C. Ainsworth and P. H. A. Sneath), Cambridge University Press, Cambridge, pp. 1–13.

Caliński, T. and Harabasz, J. (1974) A dendrite method for cluster analysis. *Communications in Statistics*, **3**, 1–27.

Carroll, J. D. and Arabie, P. (1983) INDCLUS: an individual differences generalization of the ADCLUS model and the MAPCLUS algorithm. *Psychometrika*, **48**, 157–69.

Carroll, J. D. and Arabie, P. (1998) Multidimensional scaling, in *Measurement, Judgment and Decision Making* (ed M. H. Birnbaum), Academic Press, San Diego, CA, pp. 179–250.

Carroll, J. D., Clark, L. A. and DeSarbo, W. S. (1984) The representation of three-way proximity data by single and multiple tree structure models. *Journal of Classification*, **1**, 25–74.

Carroll, J. D. and Pruzansky, S. (1975) Fitting of hierarchical tree structure (HTS) models, mixtures of HTS models, and hybrid models, via mathematical programming and alternating least squares. *Paper presented at U. S. - Japan Seminar on Theory, Methods and Applications of Multidimensional Scaling and Related Techniques*, San Diego, August 20-24, 1975.

Carroll, J. D. and Pruzansky, S. (1980) Discrete and hybrid scaling models, in *Similarity and Choice* (eds E. D. Lantermann and H. Feger), Huber, Bern, pp. 108–39.

Chand, D. R. and Kapur, S. S. (1970) An algorithm for convex polytopes. *Journal of the Association for Computing Machinery*, **17**, 78–86.

Chandon, J. L., Lemaire, J. and Pouget, J. (1980) Construction de l'ultramétrique la plus proche d'une dissimilarité au sens des moindres carrés. *Recherche Opérationelle / Operations Research*, **14**, 157–70.

Chazelle, B. (1985) Fast searching in a real algebraic manifold with applications to geometric complexity. *Lecture Notes in Computer Science*, **185**, 145–56.

Chen, Z. and Van Ness, J. W. (1994) Metric admissibility and agglomerative clustering. *Communications in Statistics – Simulation and Computation*, **23**, 833–45.

Chen, Z. and Van Ness, J. W. (1996) Space-conserving agglomerative algorithms. *Journal of Classification*, **13**, 157–68.

Cheng, R. and Milligan, G. W. (1996) Measuring the influence of individual data points in a cluster analysis. *Journal of Classification*, **13**, 315–35.

Cleveland, W. S. and McGill, M. E. (Eds) (1988) *Dynamic Graphics for Statistics*, Wadsworth & Brooks/Cole, Belmont, CA.

Cliff, N., Girard, R., Green, R. S., Kehoe, J. F. and Doherty, L. M. (1977) INTERSCAL: a TSO FORTRAN IV program for subject computer interactive multidimensional scaling. *Educational and Psychological Measurement*, **37**, 185–8.

Cole, A. J. and Wishart, D. (1970) An improved algorithm for the Jardine-Sibson method of generating overlapping clusters. *Computer Journal*, **13**, 156–63.

Constantinescu, M. and Sankoff, D. (1995) An efficient algorithm for supertrees. *Journal of Classification*, **12**, 101–12.

Cormack, R. M. (1971) A review of classification (with discussion). *Journal of the Royal Statistical Society*, **A 134**, 321–67.

Cox, T. F. and Cox, M. A. A. (1994) *Multidimensional Scaling*, Chapman & Hall, London.

Crawford, R. M. M. and Wishart, D. (1967) A rapid multivariate method for the detection and classification of groups of ecologically related species. *Journal of Ecology*, **55**, 505–24.

Crawford, R. M. M. and Wishart, D. (1968) A rapid classification and ordination method and its application to vegetation mapping. *Journal of Ecology*, **56**, 385–404.

Critchley, F. (1978) Multidimensional scaling: a short critique and a new method, in *COMPSTAT 1978* (eds L. C. A. Corsten and J. Hermans), Physica, Vienna, pp. 297–303.

Cross, G. R. and Jain, A. K. (1982) Measurement of clustering tendency, in *Proceedings of IFAC Symposium on Theory and Application of Digital Control, Volume 2*, New Delhi, pp. 24–9.

Daskin, M. S. (1995) *Network and Discrete Location Models, Algorithms, and Applications*, Wiley, New York.

Daws, J. T. (1996) The analysis of free-sorting data: beyond pairwise cooccurrences. *Journal of Classification*, **13**, 57–80.

Day, W. H. E. (1996) Complexity theory: an introduction for practitioners of classification, in *Clustering and Classification* (eds P. Arabie, L. J. Hubert and G. De Soete), World Scientific, Singapore, pp. 199–233.

Day, W. H. E. and Edelsbrunner, H. (1984) Efficient algorithms for agglomerative hierarchical clustering methods. *Journal of Classification*, **1**, 7–24.

Day, W. H. E. and Edelsbrunner, H. (1985) Investigations of proportional link linkage clustering methods. *Journal of Classification*, **2**, 239–54.

de Carvalho, F. A. T. (1998) Extension based proximities between constrained Boolean symbolic objects, in *Data Science, Classification, and Related Methods* (eds C. Hayashi, N. Ohsumi, K. Yajima, Y. Tanaka, H.-H. Bock and Y. Baba), Springer, Tokyo, pp. 370–8.

Defays, D. (1977) An efficient algorithm for a complete link method. *Computer Journal*, **20**, 364–6.

Delattre, M. and Hansen, P. (1980) Bicriterion cluster analysis. *IEEE Transactions on Pattern Analysis and Machine Intelligence*, **PAMI-2**, 277–91.

de Queiroz, A. (1993) For consensus (sometimes). *Systematic Biology*, **42**, 368–72.

DeSarbo, W. S., Carroll, J. D., Clark, L. A. and Green, P. E. (1984) Synthesized clustering: a method for amalgamating alternative clustering bases with differential weighting of variables. *Psychometrika*, **49**, 57–78.

DeSarbo, W. S. and Mahajan, V. (1984) Constrained classification: the use of a priori information in cluster analysis. *Psychometrika*, **49**, 187–215.

De Soete, G. (1984a) A least squares algorithm for fitting an ultrametric tree to a dissimilarity matrix. *Pattern Recognition Letters*, **2**, 133–7.

De Soete, G. (1984b) Ultrametric tree representations of incomplete dissimilarity data. *Journal of Classification*, **1**, 235–42.

De Soete, G. (1986) Optimal variable weighting for ultrametric and additive tree clustering. *Quality and Quantity*, **20**, 169–80.

De Soete, G. and Carroll, J. D. (1996) Tree and other network models for representing proximity data, in *Clustering and Classification* (eds P. Arabie, L. J. Hubert and G. De Soete), World Scientific, Singapore, pp. 157–97.

De Soete, G., DeSarbo, W. S. and Carroll, J. D. (1985) Optimal variable weighting for hierarchical clustering: an alternating least-squares algorithm. *Journal of Classification*, **2**, 173–92.

De Soete, G., DeSarbo, W. S., Furnas, G. W. and Carroll, J. D. (1984) Tree representations of rectangular proximity matrices, in *Trends in Mathematical Psychology* (eds. E. Degreef and J. Van Buggenhaut), Elsevier, Amsterdam, pp. 377–92.

Diday, E. (1984) Une représentation visuelle des classes empietantes: les pyramides. *Rapport de Recherche 291*, INRIA Rocquencourt.

Diday, E. (1986) Orders and overlapping clusters by pyramids, in *Multidimensional Data Analysis* (eds J. De Leeuw, W. Heiser, J. Meulman and F. Critchley), DSWO Press, Leiden, pp. 201–34.

Diday, E. (1988) The symbolic approach in clustering and related methods of data analysis: the basic choices, in *Classification and Related Methods of Data Analysis* (ed H. H. Bock), North-Holland, Amsterdam, pp. 673–83.

Diday, E. (1995) Probabilist, possibilist and belief objects for knowledge analysis. *Annals of Operations Research*, **55**, 227–76.

Diday, E. and Bertrand, P. (1986) An extension of hierarchical clustering: the pyramidal presentation, in *Pattern Recognition in Practice II* (eds E. S. Gelsema and L. N. Kanal), North-Holland, Amsterdam, pp. 411–23.

Diday, E. and Govaert, G. (1977) Classification automatique avec distances adaptatives. *RAIRO Informatique / Computer Sciences*, **11**, 329–49.

Diehr, G. (1985) Evaluation of a branch and bound algorithm for clustering. *SIAM Journal on Scientific and Statistical Computing*, **6**, 268–84.

Diggle, P. J. (1983) *Statistical Analysis of Spatial Point Patterns*, Academic Press, London.

Dobkin, D. and Lipton, R. J. (1976) Multidimensional searching problems. *SIAM Journal on Computing*, **5**, 181–6.

Dubien, J. L. and Warde, W. D. (1979) A mathematical comparison of the members of an infinite family of agglomerative clustering algorithms. *Canadian Journal of Statistics*, **7**, 29–38.

Duda, R. O. and Hart, P. E. (1973) *Pattern Classification and Scene Analysis*, Wiley, New York.

Dunn, J. C. (1974) A fuzzy relative of the ISODATA process and its use in detecting compact well-separated clusters. *Journal of Cybernetics*, **3(3)**, 32–57.

Durand, C. and Fichet, B. (1988) One-to-one correspondences in pyramidal representation: a unified approach, in *Classification and Related Methods of Data Analysis* (ed H. H. Bock), North-Holland, Amsterdam, pp. 85–90.

Eckart, C. and Young, G. (1936) The approximation of one matrix by another of lower rank. *Psychometrika*, 1, 211–8.

Edelsbrunner, H. (1987) *Algorithms in Combinatorial Geometry*, Springer, Berlin.

Edwards, A. W. F. and Cavalli-Sforza, L. L. (1965) A method for cluster analysis. *Biometrics*, 21, 362–75.

Erlenkotter, D. (1978) A dual-based procedure for uncapacitated facility location. *Operations Research*, 26, 992–1009.

Fayyad, U. M., Piatetsky-Shapiro, G., Smyth, P. and Uthurusamy, R. (Eds) (1996) *Advances in Knowledge Discovery and Data Mining*, AAAI Press / MIT Press, Menlo Park, CA.

Feng, C. and Michie, D. (1994) Machine learning of rules and trees, in *Machine Learning, Neural and Statistical Classification* (eds D. Michie, D. J. Spiegelhalter and C. C. Taylor), Ellis Horwood, Hemel Hempstead, pp. 50–83.

Ferligoj, A. and Batagelj, V. (1982) Clustering with relational constraint. *Psychometrika*, 47, 413–26.

Ferligoj, A. and Batagelj, V. (1983) Some types of clustering with relational constraints. *Psychometrika*, 48, 541–52.

Fichet, B. and Le Calvé, G. (1984) Structure géometrique des principaux indices de dissimilarité sur signes de présence-absence. *Statistiques et Analyse de Données*, 9, 11–44.

Finden, C. R. and Gordon, A. D. (1985) Obtaining common pruned trees. *Journal of Classification*, 2, 255–76.

Fisher, D. H. (1987) Knowledge acquisition via incremental conceptual clustering. *Machine Learning*, 2, 139–72.

Fisher, D. (1996) Iterative optimization and simplification of hierarchical clusterings. *Journal of Artificial Intelligence Research*, 4, 147–80.

Fisher, L. and Van Ness, J. W. (1971) Admissible clustering procedures. *Biometrika*, 58, 91–104.

Fisher, W. D. (1958) On grouping for maximum homogeneity. *Journal of the American Statistical Association*, 53, 789–98.

Fisherkeller, M. A., Friedman, J. H. and Tukey, J. W. (1988) PRIM-9: an interactive multidimensional data display and analysis system, in *The Collected Works of John W. Tukey, Volume V Graphics: 1965–1985* (ed W. S. Cleveland), Wadsworth & Brooks/Cole, Pacific Grove, CA, pp. 307–27.

Florek, K., Lukaszewicz, J., Perkal, J., Steinhaus, H. and Zubrzycki, S. (1951) Sur la liaison et la division des points d'un ensemble fini. *Colloquium Mathematicum*, 2, 282–5.

Fowlkes, E. B., Gnanadesikan, R. and Kettenring, J. R. (1988) Variable selection in clustering. *Journal of Classification*, **5**, 205–28.

Frank, O. and Svensson, K. (1981) On probability distributions of single-linkage dendrograms. *Journal of Statistical Computation and Simulation*, **12**, 121–31.

Frawley, W. J., Piatetsky-Shapiro, G. and Matheus, C. J. (1992) Knowledge discovery in databases: an overview. *Artificial Intelligence Magazine*, **13(3)**, 57–70.

Friedman, H. P. and Rubin, J. (1967) On some invariant criteria for grouping data. *Journal of the American Statistical Association*, **62**, 1159–78.

Friedman, J. H. and Rafsky, L. C. (1979) Multivariate generalizations of the Wald-Wolfowitz and Smirnov two-sample tests. *Annals of Statistics*, **7**, 697–717.

Friedman, J. H. and Tukey, J. W. (1974) A projection pursuit algorithm for exploratory data analysis. *IEEE Transactions on Computers*, **C-23**, 881–90.

Fukada, Y. (1980) Spatial clustering procedures for region analysis. *Pattern Recognition*, **12**, 395–403.

Furnas, G. W. (1984) The generation of random, binary unordered trees. *Journal of Classification*, **1**, 187–233.

Gabriel, K. R. (1971) The biplot graphic display of matrices with application to principal component analysis. *Biometrika*, **58**, 453–67.

Gabriel, K. R. and Sokal, R. R. (1969) A new statistical approach to geographical variation analysis. *Systematic Zoology*, **18**, 259–78.

Gabriel, K. R. and Zamir, S. (1979) Lower rank approximation of matrices by least squares with any choice of weights. *Technometrics*, **21**, 489–98.

Gale, N., Halperin, W. C. and Costanzo, C. M. (1984) Unclassed matrix shading and optimal ordering in hierarchical cluster analysis. *Journal of Classification*, **1**, 75–92 (Erratum: **1**, 289).

Garey, M. R. and Johnson, D. S. (1979) *Computers and Intractability: A Guide to the Theory of NP-Completeness*, W. H. Freeman, San Francisco.

Garfinkel, R. S., Neebe, A.W. and Rao, M. R. (1974) An algorithm for the $m$-median plant location problem. *Transportation Science*, **8**, 217–36.

Gaul, W. and Schader, M. (1994) Pyramidal classification based on incomplete dissimilarity data. *Journal of Classification*, **11**, 171–93.

Gettler Summa, M., Périnel, E. and Ferraris, J. (1994) Automatic aid to symbolic cluster interpretation, in *New Approaches in Classification and Data Analysis* (eds E. Diday, Y. Lechevallier, M. Schader, P. Bertrand and B. Burtschy), Springer, Berlin, pp. 405–13.

Ghashgai, E., Stinebrickner, R. and Suters, W. H. (1989) A family of consensus dendrogram methods. *Paper presented at the Second Conference of the International Federation of Classification Societies,* Charlottesville, VA, June 27-30, 1989.

Gifi, A. (1990) *Nonlinear Multivariate Analysis,* John Wiley & Sons, Chichester.

Gill, D. (1970) Application of a statistical zonation method to reservoir evaluation and digitized-log analysis. *American Association of Petroleum Geologists Bulletin,* **54**, 719-29.

Gilmore, P. C. and Gomory, R. E. (1961) A linear programming approach to the cutting stock problem. *Operations Research,* **9**, 849-59.

Gilmour, J. S. L. (1937) A taxonomic problem. *Nature,* **139**, 1040-2.

Gnanadesikan, R., Kettenring, J. R. and Landwehr, J. M. (1977) Interpreting and assessing the results of cluster analyses. *Bulletin of the International Statistical Institute,* **47 (2)**, 451-63.

Goddard, W., Kubicka, E., Kubicki, G. and McMorris, F. R. (1994) The agreement metric for labeled binary trees. *Mathematical Biosciences,* **123**, 215-26.

Goddard, W., Kubicka, E., Kubicki, G. and McMorris, F. R. (1995) Agreement subtrees, metric and consensus for labeled binary trees, in *Partitioning Data Sets* (eds I. J. Cox, P. Hansen and B. Julesz), DIMACS Series in Discrete Mathematics and Theoretical Computer Science, Volume 19, American Mathematical Society, Providence, RI, pp. 97-104.

Good, I. J. (1965) Categorization of classification, in *Mathematics and Computer Science in Biology and Medicine,* HMSO, London, pp. 115-28.

Goodman, L. A. and Kruskal, W. H. (1954) Measures of association for cross-classifications. *Journal of the American Statistical Association,* **49**, 732-64.

Gordon, A. D. (1973) Classification in the presence of constraints. *Biometrics,* **29**, 821-7.

Gordon, A. D. (1979) A measure of the agreement between rankings. *Biometrika,* **66**, 7-15.

Gordon, A. D. (1980) On the assessment and comparison of classifications, in *Analyse de Données et Informatique* (ed R. Tomassone), INRIA, Le Chesnay, pp. 149-60.

Gordon, A. D. (1981) *Classification: Methods for the Exploratory Analysis of Multivariate Data,* Chapman & Hall, London.

Gordon, A. D. (1986a) Links between clustering and assignment procedures, in *COMPSTAT 1986* (eds F. De Antoni, N. Lauro and A. Rizzi), Physica, Heidelberg, pp. 149-56.

Gordon, A. D. (1986b) Consensus supertrees: the synthesis of rooted trees containing overlapping sets of labeled leaves. *Journal of Classification,* **3**, 335-48.

Gordon, A. D. (1987) Parsimonious trees. *Journal of Classification*, **4**, 85–101.

Gordon, A. D. (1990) Constructing dissimilarity measures. *Journal of Classification*, **7**, 257–69.

Gordon, A. D. (1994) Identifying genuine clusters in a classification. *Computational Statistics & Data Analysis*, **18**, 561–81.

Gordon, A. D. (1996a) Hierarchical classification, in *Clustering and Classification* (eds P. Arabie, L. J. Hubert and G. De Soete), World Scientific, Singapore, pp. 65–121.

Gordon, A. D. (1996b) A survey of constrained classification. *Computational Statistics & Data Analysis*, **21**, 17–29.

Gordon, A. D. (1996c) Null models in cluster validation, in *From Data to Knowledge: Theoretical and Practical Aspects of Classification, Data Analysis, and Knowledge Organization* (eds W. Gaul and D. Pfeifer), Springer, Berlin, pp. 32–44.

Gordon, A. D. (1998) Cluster validation, in *Data Science, Classification, and Related Methods* (eds C. Hayashi, N. Ohsumi, K. Yajima, Y. Tanaka, H.-H. Bock and Y. Baba), Springer, Tokyo, pp. 22–39.

Gordon, A. D. and Birks, H. J. B. (1972) Numerical methods in Quaternary palaeoecology I. Zonation of pollen diagrams. *New Phytologist*, **71**, 961–79.

Gordon, A. D. and De Cata, A. (1988) Stability and influence in sum of squares clustering. *Metron*, **46**, 347–60.

Gordon, A. D. and Vichi, M. (1998) Partitions of partitions. *Journal of Classification*, **15**, 265–85.

Gostick, R. W. (1979) Software and algorithms for the distributed-array processors. *ICL Technical Journal*, **2**, 116–35.

Gowda, K. C. and Diday, E. (1991) Symbolic clustering using a new dissimilarity measure. *Pattern Recognition*, **24**, 567–78.

Gowda, K. C. and Diday, E. (1994) Symbolic clustering algorithms using similarity and dissimilarity measures, in *New Approaches in Classification and Data Analysis* (eds E. Diday, Y. Lechevallier, M. Schader, P. Bertrand and B. Burtschy), Springer, Berlin, pp. 414–22.

Gower, J. C. (1966) Some distance properties of latent root and vector methods used in multivariate analysis. *Biometrika*, **53**, 325–38.

Gower, J. C. (1967) A comparison of some methods of cluster analysis. *Biometrics*, **23**, 623–38.

Gower, J. C. (1971a) A general coefficient of similarity and some of its properties. *Biometrics*, **27**, 857–74.

Gower, J. C. (1971b) Discussion of paper by R. M. Cormack. *Journal of the Royal Statistical Society*, **A 134**, 360–5.

Gower, J. C. (1974) Maximal predictive classification. *Biometrics*, **30**, 643–54.

Gower, J. C. (1977) The analysis of asymmetry and orthogonality, in

*Recent Developments in Statistics* (eds J. R. Barra, F. Brodeau, G. Romier and B. Van Cutsem), North-Holland, Amsterdam, pp. 109–23.

Gower, J. C. (1985) Measures of similarity, dissimilarity and distance, in *Encyclopedia of Statistical Sciences, Volume 5* (eds S. Kotz, N. L. Johnson and C. B. Read), Wiley, New York, pp. 397–405.

Gower, J. C. and Banfield, C. F. (1975) Goodness-of-fit criteria for hierarchical classification and their empirical distributions, in *Proceedings of the 8th International Biometric Conference* (eds L. C. A. Corsten and T. Postelnicu), Constanţa, Romania, pp. 347–61.

Gower, J. C. and Hand, D. J. (1996) *Biplots*, Chapman & Hall, London.

Gower, J. C. and Legendre, P. (1986) Metric and Euclidean properties of dissimilarity coefficients. *Journal of Classification*, **3**, 5–48.

Graef, J. and Spence, I. (1979) Using distance information in the design of large multidimensional scaling experiments. *Psychological Bulletin*, **86**, 60–6.

Graham, R. L. and Hell, P. (1985) On the history of the minimum spanning tree problem. *Annals of the History of Computing*, **7**, 43–57.

Green, R. S. and Bentler, P. M. (1979) Improving the efficiency and effectiveness of interactively selected MDS data designs. *Psychometrika*, **44**, 115–9.

Greenacre, M. J. (1984) *Theory and Applications of Correspondence Analysis*, Academic Press, London.

Greig-Smith, P. (1964) *Quantitative Plant Ecology*, 2nd edn, Butterworths, London.

Grötschel, M. and Wakabayashi, Y. (1989) A cutting plane algorithm for a clustering problem. *Mathematical Programming*, **45**, 59–96.

Guénoche, A., Hansen, P. and Jaumard, B. (1991) Efficient algorithms for divisive hierarchical clustering with the diameter criterion. *Journal of Classification*, **8**, 5–30.

Hanjoul, P. and Peeters, D. (1985) A comparison of two dual-based procedures for solving the *p*-median problem. *European Journal of Operational Research*, **20**, 387–96.

Hansen, P. and Delattre, M. (1978) Complete-link cluster analysis by graph coloring. *Journal of the American Statistical Association*, **73**, 397–403.

Hansen, P. and Jaumard, B. (1987) Minimum sum of diameters clustering. *Journal of Classification*, **4**, 215–26.

Hansen, P. and Jaumard, B. (1997) Cluster analysis and mathematical programming. *Mathematical Programming*, **79**, 191–215.

Hansen, P., Jaumard, B. and Mladenovic, N. (1995) How to choose *K* entities among *N*, in *Partitioning Data Sets* (eds I. J. Cox, P. Hansen and B. Julesz), DIMACS Series in Discrete Mathematics and Theoretical Computer Science, Volume 19, American Mathematical Society,

Providence, RI, pp. 105–16.

Hansen, P., Jaumard, B. and Sanlaville, E. (1994) Partitioning problems in cluster analysis: a review of mathematical programming approaches, in *New Approaches in Classification and Data Analysis* (eds E. Diday, Y. Lechevallier, M. Schader, P. Bertrand and B. Burtschy), Springer, Berlin, pp. 228–40.

Hansen, P., Jaumard, B. and Simeone, B. (1996) Espaliers: a generalization of dendrograms. *Journal of Classification*, **13**, 107–27.

Harper, C. W., Jr. (1978) Groupings by locality in community ecology and paleoecology: tests of significance. *Lethaia*, **11**, 251–7.

Hart, G. (1983) The occurrence of multiple UPGMA dendrograms, in *Numerical Taxonomy* (ed J. Felsenstein), Springer, Berlin, pp. 254–8.

Hartigan, J. A. (1967) Representation of similarity matrices by trees. *Journal of the American Statistical Association*, **62**, 1140–58.

Hartigan, J. A. (1975) *Clustering Algorithms*, Wiley, New York.

Hartigan, J. A. (1988) The span test for unimodality, in *Classification and Related Methods of Data Analysis* (ed H. H. Bock), North-Holland, Amsterdam, pp. 229–36.

Hartigan, J. A. and Mohanty, S. (1992) The runt test for multimodality. *Journal of Classification*, **9**, 63–70.

Hartigan, J. A. and Wong, M. A. (1979) Algorithm AS 136. A $k$-means clustering algorithm. *Applied Statistics*, **28**, 100–8.

Hathaway, R. J. and Bezdek, J. C. (1988) Recent convergence results for the fuzzy $c$-means clustering algorithms. *Journal of Classification*, **5**, 237–47.

Hawkins, D. M. and Merriam, D. F. (1973) Optimal zonation of digitized sequential data. *Mathematical Geology*, **5**, 389–95.

Hayashi, C. (1972) Two dimensional quantification based on the measure of dissimilarity among three elements. *Annals of the Institute of Statistical Mathematics*, **24**, 251–7.

Hébrail, G. (1998) The SODAS project: a software for symbolic data analysis, in *Data Science, Classification, and Related Methods* (eds C. Hayashi, N. Ohsumi, K. Yajima, Y. Tanaka, H.-H. Bock and Y. Baba), Springer, Tokyo, pp. 387–93.

Ho, T. B., Diday, E. and Gettler-Summa, M. (1988) Generating rules for expert systems from observations. *Pattern Recognition Letters*, **7**, 265–71.

Hodson, F. R., Sneath, P. H. A. and Doran, J. E. (1966) Some experiments in the numerical analysis of archaeological data. *Biometrika*, **53**, 311–24.

Hoffman, R. and Jain, A. K. (1983) A test of randomness based on the minimal spanning tree. *Pattern Recognition Letters*, **1**, 175–80.

Hope, A. C. A. (1968) A simplified Monte Carlo significance test procedure. *Journal of the Royal Statistical Society*, **B 30**, 582–98.

Horowitz, S. L. and Pavlidis, T. (1976) Picture segmentation by a tree traversal algorithm. *Journal of the Association for Computing Machinery*, **23**, 368–88.

Hotelling, H. (1933) Analysis of a complex of statistical variables into principal components. *Journal of Educational Psychology*, **24**, 417–41.

Howe, S. E. (1979) *Estimating Regions and Clustering Spatial Data: Analysis and Implementation of Methods Using the Voronoi Diagram*, Ph.D. thesis, Brown University.

Huang, J. S. and Tseng, D. H. (1988) Statistical theory of edge detection. *Computer Vision, Graphics, and Image Processing*, **43**, 337–46.

Hubac, J. M. (1964) Application de la taxonomie de Wraclaw (technique des dendrites) à quelques populations du *Campanula rotundifolia* L., s. l., et utilisation de cette technique pour l'établissement des clés de détermination. *Bulletin de la Société Botanique de France*, **111**, 331–46.

Hubálek, Z. (1982) Coefficients of association and similarity, based on binary (presence-absence) data: an evaluation. *Biological Reviews of the Cambridge Philosophical Society*, **57**, 669–89.

Huber, P. J. (1985) Projection pursuit (with discussion). *Annals of Statistics*, **13**, 435–525.

Hubert, L. (1973a) Min and max hierarchical clustering using asymmetric similarity measures. *Psychometrika*, **38**, 63–72.

Hubert, L. (1973b) Monotone invariant clustering procedures. *Psychometrika*, **38**, 47–62.

Hubert, L. J. (1974a) Some applications of graph theory to clustering. *Psychometrika*, **39**, 283–309.

Hubert, L. (1974b) Approximate evaluation techniques for the single-link and complete-link hierarchical clustering procedures. *Journal of the American Statistical Association*, **69**, 698–704.

Hubert, L. J. (1987) *Assignment Methods in Combinatorial Data Analysis*, Marcel Dekker, New York.

Hubert, L. and Arabie, P. (1985) Comparing partitions. *Journal of Classification*, **2**, 193–218.

Hubert, L. and Arabie, P. (1986) Unidimensional scaling and combinatorial optimization, in *Multidimensional Data Analysis* (eds J. De Leeuw, W. Heiser, J. Meulman and F. Critchley), DSWO Press, Leiden, pp. 181–96.

Hubert, L. and Arabie, P. (1988) Relying on necessary conditions for optimization: unidimensional scaling and some extensions, in *Classification and Related Methods of Data Analysis* (ed H. H. Bock), North-Holland, Amsterdam, pp. 463–72.

Hubert, L. and Arabie, P. (1995) Iterative projection strategies for the least-squares fitting of tree structures to proximity data. *British Journal of Mathematical and Statistical Psychology*, **48**, 281–317.

Hubert, L., Arabie, P. and Meulman, J. (1997) Hierarchical clustering and the construction of (optimal) ultrametrics using $L_p$-norms,
in $L_1$-Statistical Procedures and Related Topics (ed Y. Dodge), Lecture Notes – Monograph Series, Volume 31, Institute of Mathematical
Statistics, Hayward, CA, pp. 457–72.

Hwang, K. and Briggs, F. A. (1984) Computer Architecture and Parallel
Processing, McGraw-Hill, New York.

Ichino, M. and Yaguchi, H. (1994) Generalized Minkowski metrics for
mixed feature-type data analysis. IEEE Transactions on Systems,
Man, and Cybernetics, 24, 698–708.

Intrator, N. (1992) Feature extraction using an unsupervised neural network. Neural Computation, 4, 98–107.

Intrator, N. and Cooper, L. N. (1992) Objective function formulation of
the BCM theory of visual cortical plasticity: statistical connections,
stability conditions. Neural Networks, 5, 3–17.

Ismail, M. A. and Kamel, M. S. (1989) Multidimensional data clustering
utilizing hybrid search strategies. Pattern Recognition, 22, 75–89.

Jain, A. K. and Dubes, R. C. (1988) Algorithms for Clustering Data,
Prentice-Hall, Englewood Cliffs, NJ.

Jambu, M. (1978) Classification Automatique pour l'Analyse des
Données, North-Holland, Amsterdam.

Jardine, C. J., Jardine, N. and Sibson, R. (1967) The structure and
construction of taxonomic hierarchies. Mathematical Biosciences, 1,
173–9.

Jardine, N. (1969) Towards a general theory of clustering. Biometrics,
25, 609–10.

Jardine, N. and Sibson, R. (1968) The construction of hierarchic and
non-hierarchic classifications. Computer Journal, 11, 177–84.

Jardine, N. and Sibson, R. (1971) Mathematical Taxonomy, Wiley, London.

Jensen, R. E. (1969) A dynamic programming algorithm for cluster analysis. Operations Research, 17, 1034–57.

Johnson, R. M. (1963) On a theorem stated by Eckart and Young. Psychometrika, 28, 259–63.

Johnson, S. C. (1967) Hierarchical clustering schemes. Psychometrika,
32, 241–54.

Jolliffe, I. T. (1986) Principal Component Analysis, Springer, New York.

Jolliffe, I. T., Jones, B. and Morgan, B. J. T. (1988) Stability and influence in cluster analysis, in Data Analysis and Informatics, V (ed E.
Diday), North-Holland, Amsterdam, pp. 507–14.

Joly, S. and Le Calvé, G. (1995) Three-way distances. Journal of Classification, 12, 191–205.

Jones, M. C. and Sibson, R. (1987) What is projection pursuit? (with
discussion). Journal of the Royal Statistical Society, A 150, 1–36.

Juan, J. (1982) Programme de classification hiérarchique par l'algorithme de la recherche en chaîne des voisins réciproques. *Les Cahiers de l'Analyse des Données*, **7**, 219–25.

Kaufman, L. and Rousseeuw, P. J. (1990) *Finding Groups in Data: An Introduction to Cluster Analysis*, Wiley, New York.

Kelly, F. P. and Ripley, B. D. (1976) A note on Strauss's model for clustering. *Biometrika*, **63**, 357–60.

Kendall, D. G. (1971) Seriation from abundance matrices, in *Mathematics in the Archaeological and Historical Sciences* (eds F. R. Hodson, D. G. Kendall and P. Tǎutu), Edinburgh University Press, Edinburgh, pp. 215–52.

Kendall, M. G. (1966) Discrimination and classification, in *Multivariate Analysis* (ed P. R. Krishnaiah), Academic Press, New York, pp. 165–85.

Kendrick, W. B. (1965) Complexity and dependence in computer taxonomy. *Taxon*, **14**, 141–54.

Kendrick, W. B. and Proctor, J. R. (1964) Computer taxonomy in the fungi imperfecti. *Canadian Journal of Botany*, **42**, 65–8.

Kiers, H. A. L. and Takane, Y. (1994) A generalization of GIPSCAL for the analysis of nonsymmetric data. *Journal of Classification*, **11**, 79–99.

Kirkpatrick, S., Gelatt, C. D., Jr. and Vecchi, M. P. (1983) Optimization by simulated annealing. *Science*, **220**, 671–80.

Klastorin, T. D. and Watts, C. A. (1981) The determination of alternative hospital classifications. *Health Services Research*, **16**, 205–20.

Klein, G. and Aronson, J. E. (1991) Optimal clustering: a model and method. *Naval Research Logistics*, **38**, 447–61.

Klein, R. W. and Dubes, R. C. (1989) Experiments in projection and clustering by simulated annealing. *Pattern Recognition*, **22**, 213–20.

Kleiner, B. and Hartigan, J. A. (1981) Representing points in many dimensions by trees and castles (with discussion). *Journal of the American Statistical Association*, **76**, 260–76.

Kohonen, T. (1982) Self-organized formation of topologically correct feature maps. *Biological Cybernetics*, **43**, 59–69.

Kohonen, T. (1990) The self-organizing map. *Proceedings of the IEEE*, **78**, 1464–80.

Koontz, W. L. G., Narendra, P. M. and Fukunaga, K. (1975) A branch and bound clustering algorithm. *IEEE Transactions on Computers*, **C-24**, 908–15.

Křivánek, M. (1986) On the computational complexity of clustering, in *Data Analysis and Informatics, IV* (eds E. Diday, Y. Escoufier, L. Lebart, J. Pagès, Y. Schektman and R. Tomassone), North-Holland, Amsterdam, pp. 89–96.

Křivánek, M. and Morávek, J. (1986) NP- hard problems in hierarchical-tree clustering. *Acta Informatica*, **23**, 311–23.

Kruskal, J. B. (1956) On the shortest spanning subtree of a graph and the traveling salesman problem. *Proceedings of the American Mathematical Society*, **7**, 48–50.

Kruskal, J. B. (1964a) Multidimensional scaling by optimizing goodness of fit to a nonmetric hypothesis. *Psychometrika*, **29**, 1–27.

Kruskal, J. B. (1964b) Nonmetric multidimensional scaling: a numerical method. *Psychometrika*, **29**, 115–29.

Kruskal, J. B. (1969) Towards a practical method which helps uncover the structure of a set of multivariate observations by finding the linear transformation which optimizes a new 'index of condensation', in *Statistical Computation* (eds R. C. Milton and J. A. Nelder), Academic Press, New York, pp. 427–40.

Kruskal, J. B. (1972) Linear transformation of multivariate data to reveal clustering, in *Multidimensional Scaling. Theory and Applications in the Behavioral Sciences Volume 1 Theory* (eds R. N. Shepard, A. K. Romney and S. B. Nerlove), Seminar Press, New York, pp. 179–91.

Kruskal, J. B. and Landwehr, J. M. (1983) Icicle plots: better displays for hierarchical clustering. *American Statistician*, **37**, 162–8.

Kruskal, J. B. and Wish, M. (1978) *Multidimensional Scaling*, Sage Publications, Beverly Hills.

Krzanowski, W. J. and Marriott, F. H. C. (1994) *Multivariate Analysis, Part 1. Distributions, Ordination and Inference*, Arnold, London

Krzanowski, W. J. and Marriott, F. H. C. (1995) *Multivariate Analysis, Part 2. Classification, Covariance Structures and Repeated Measurements*, Arnold, London.

Lance, G. N. and Williams, W. T. (1966a) Computer programs for hierarchical polythetic classification ('similarity analyses'). *Computer Journal*, **9**, 60–4.

Lance, G. N. and Williams, W. T. (1966b) A generalised sorting strategy for computer classifications. *Nature*, **212**, 218.

Lance, G. N. and Williams, W. T. (1967) A general theory of classificatory sorting strategies I. Hierarchical systems. *Computer Journal*, **9**, 373–80.

Lance, G. N. and Williams, W. T. (1968) Note on a new information-statistic classificatory program. *Computer Journal*, **11**, 195.

Lankford, P. M. (1969) Regionalization: theory and alternative algorithms. *Geographical Analysis*, **1**, 196–212.

Lapointe, F.-J. (1998) How to validate phylogenetic trees? A stepwise procedure, in *Data Science, Classification, and Related Methods* (eds C. Hayashi, N. Ohsumi, K. Yajima, Y. Tanaka, H.-H. Bock and Y. Baba), Springer, Tokyo, pp. 71–88.

Lapointe, F.-J. and Cucumel, G. (1991) The average consensus. *Paper presented at the Third Conference of the International Federation of*

*Classification Societies*, Edinburgh, August 6-9, 1991.

Lapointe, F.-J. and Cucumel, G. (1995) The average consensus procedure: combination of weighted trees containing identical or overlapping sets of taxa. *Systematic Zoology*, **46**, 306–12.

Lapointe, F.-J. and Legendre, P. (1990) A statistical framework to test the consensus of two nested classifications. *Systematic Zoology*, **39**, 1–13.

Lapointe, F.-J. and Legendre, P. (1991) The generation of random ultrametric matrices representing dendrograms. *Journal of Classification*, **8**, 177–200.

Lapointe, F.-J. and Legendre, P. (1994) A classification of pure malt Scotch whiskies. *Applied Statistics*, **43**, 237–57.

Lau, K., Leung, P. L. and Tse, K. (1998) A nonlinear programming approach to metric unidimensional scaling. *Journal of Classification*, **15**, 3–14.

Lebart, L. (1978) Programme d'agrégation avec contraintes. *Les Cahiers de l'Analyse des Données*, **3**, 275–87.

Le Cam, L. (1970) On the assumptions used to prove asymptotic normality of maximum likelihood estimates. *Annals of Mathematical Statistics*, **41**, 802–28.

Leclerc, B. (1998) Consensus of classifications: the case of trees, in *Advances in Data Science and Classification* (eds A. Rizzi, M. Vichi and H.-H. Bock), Springer, Berlin, pp. 81–90.

Leclerc, B. and Cucumel, G. (1987) Consensus en classification: une revue bibliographique. *Mathématiques et Sciences Humaines*, **100**, 109–28.

Lefkovitch, L. P. (1978) Cluster generation and grouping using mathematical programming. *Mathematical Biosciences*, **41**, 91–110.

Lefkovitch, L. P. (1980) Conditional clustering. *Biometrics*, **36**, 43–58.

Lerman, I. C. (1970) *Les Bases de la Classification Automatique*, Gauthier-Villars, Paris.

Lerman, I. C. (1980) Combinatorial analysis in the statistical treatment of behavioral data. *Quality and Quantity*, **14**, 431–69.

Lerman, I. C. (1981) *Classification et Analyse Ordinale des Données*, Dunod, Paris.

Li, X. (1990) Parallel algorithms for hierarchical clustering and cluster validity. *IEEE Transactions on Pattern Analysis and Machine Intelligence*, **12**, 1088–92.

Li, X. and Fang, Z. (1989) Parallel clustering algorithms. *Parallel Computing*, **11**, 275–90.

Linde, Y., Buzo, A. and Gray, R. M. (1980) An algorithm for vector quantizer design. *IEEE Transactions on Communications*, **COM-28**, 84–95.

Ling, R. F. (1972) On the theory and construction of k-clusters. *Com-*

*puter Journal*, **15**, 326–32.

Ling, R. F. (1973a) A probability theory for cluster analysis. *Journal of the American Statistical Association*, **68**, 159–64.

Ling, R. F. (1973b) The expected number of components in random linear graphs. *Annals of Probability*, **1**, 876–81.

Ling, R. F. (1975) An exact probability distribution on the connectivity of random graphs. *Journal of Mathematical Psychology*, **12**, 90–8.

Ling, R. F. and Killough, G. G. (1976) Probability tables for cluster analysis based on a theory of random graphs. *Journal of the American Statistical Association*, **71**, 293–300.

Lingoes, J. C. (1971) Some boundary conditions for a monotone analysis of symmetric matrices. *Psychometrika*, **36**, 195–203.

Lingoes, J. C. and Roskam, E. E. (1973) A mathematical and empirical analysis of two multidimensional scaling algorithms. *Supplement to Psychometrika*, **38**.

Little, R. J. A. and Rubin, D. B. (1987) *Statistical Analysis with Missing Data*, Wiley, New York.

Lloyd, S. P. (1982) Least squares quantization in PCM. *IEEE Transactions on Information Theory*, **IT-28**, 129–37.

McCammon, R. B. (1968) The dendrograph: a new tool for correlation. *Bulletin of the Geological Society of America*, **79**, 1663–70.

McIntyre, R. M. and Blashfield, R. K. (1980) A nearest-centroid technique for evaluating the minimum-variance clustering procedure. *Multivariate Behavioral Research*, **15**, 225–38.

McMorris, F. R., Meronk, D. B. and Neumann, D. A. (1983) A view of some consensus methods for trees, in *Numerical Taxonomy* (ed J. Felsenstein), Springer, Berlin, pp. 122–6.

MacQueen, J. (1967) Some methods for classification and analysis of multivariate observations. *Proceedings of the 5th Berkeley Symposium on Mathematical Statistics and Probability*, **1**, 281–97.

McQuitty, L. L. (1960) Hierarchical linkage analysis for the isolation of types. *Educational and Psychological Measurement*, **20**, 55–67.

McQuitty, L. L. (1963) Rank order typal analysis. *Educational and Psychological Measurement*, **23**, 55–61.

McQuitty, L. L. (1966) Similarity analysis by reciprocal pairs for discrete and continuous data. *Educational and Psychological Measurement*, **26**, 825–31.

McQuitty, L. L. (1967) A mutual development of some typological theories and pattern-analytical methods. *Educational and Psychological Measurement*, **27**, 21–46.

Mahajan, V. and Jain, A. K. (1978) An approach to normative segmentation. *Journal of Marketing Research*, **15**, 338–45.

Mann, H. B. and Whitney, D. R. (1947) On a test of whether one of two random variables is stochastically larger than the other. *Annals of Mathematical Statistics*, **18**, 50–60.

Mao, J. and Jain, A. K. (1995) Artificial neural networks for feature extraction and multivariate data projection. *IEEE Transactions on Neural Networks*, **6**, 296–317.

Maravalle, M. and Simeone, B. (1995) A spanning tree heuristic for regional clustering. *Communications in Statistics – Theory and Methods*, **24**, 625–39.

Marcotorchino, F. and Michaud, P. (1982) Agrégation de similarités en classification automatique. *Revue de Statistique Appliquée*, **30**, 21–44.

Mardia, K. V. (1978) Some properties of classical multi-dimensional scaling. *Communications in Statistics – Theory and Methods*, **A7**, 1233–41.

Margush, T. and McMorris, F. R. (1981) Consensus $n$-trees. *Bulletin of Mathematical Biology*, **43**, 239–44.

Marriott, F. H. C. (1975) Separating mixtures of normal distributions. *Biometrics*, **31**, 767–9.

Marriott, F. H. C. (1982) Optimization methods of cluster analysis. *Biometrika*, **69**, 417–21.

Massart, D. L., Plastria, F. and Kaufman, L. (1983) Non-hierarchical clustering with MASLOC. *Pattern Recognition*, **16**, 507–16.

Matula, D. W. and Sokal, R. R. (1980) Properties of Gabriel graphs relevant to geographic variation research and the clustering of points in the plane. *Geographical Analysis*, **12**, 205–22.

Matusita, K. (1956) Decision rule, based on the distance, for the classification problem. *Annals of the Institute of Statistical Mathematics*, **8**, 67–77.

Matusita, K. (1967) Classification based on distance in multivariate Gaussian cases. *Proceedings of the 5th Berkeley Symposium on Mathematical Statistics and Probability*, **1**, 299–304.

Mehringer, P. J., Arno, S. F. and Petersen, K. L. (1977) Postglacial history of Lost Trail Pass Bog, Bitterroot Mountains, Montana. *Arctic and Alpine Research*, **9**, 345–68.

Michalski, R. S., Carbonell, J. G. and Mitchell, T. M. (Eds) (1983) *Machine Learning: An Artificial Intelligence Approach*, Tioga Publishing Company, Palo Alto, CA.

Michalski, R. S. and Stepp, R. E. (1983) Automated construction of classifications: conceptual clustering versus numerical taxonomy. *IEEE Transactions on Pattern Analysis and Machine Intelligence*, **PAMI-5**, 396–410.

Miller, G. A. and Nicely, P. E. (1955) An analysis of perceptual confusions among some English consonants. *Journal of the Acoustical Society of America*, **27**, 338–52.

Milligan, G. W. (1979) Ultrametric hierarchical clustering algorithms. *Psychometrika*, **44**, 343–6.

Milligan, G. W. (1980) An examination of the effect of six types of error perturbation on fifteen clustering algorithms. *Psychometrika*, **45**, 325–42.

Milligan, G. W. (1996) Clustering validation: results and implications for applied analyses, in *Clustering and Classification* (eds P. Arabie, L. J. Hubert and G. De Soete), World Scientific, Singapore, pp. 341–75.

Milligan, G. W. and Cooper, M. C. (1985) An examination of procedures for determining the number of clusters in a data set. *Psychometrika*, **50**, 159–79.

Milligan, G. W. and Cooper, M. C. (1986) A study of the comparability of external criteria for hierarchical cluster analysis. *Multivariate Behavioral Research*, **21**, 441–58.

Milligan, G. W. and Cooper, M. C. (1988) A study of standardization of variables in cluster analysis. *Journal of Classification*, **5**, 181–204.

Milligan, G. W. and Mahajan, V. (1980) A note on procedures for testing the quality of a clustering of a set of objects. *Decision Sciences*, **11**, 669–77.

Mirkin, B. (1987) Additive clustering and qualitative factor analysis methods for similarity matrices. *Journal of Classification*, **4**, 7–31. Erratum **6**, 271–2 (1989).

Mirkin, B. (1996) *Mathematical Classification and Clustering*, Kluwer, Dordrecht.

Miyamoto, S. and Agusta, Y. (1995) An efficient algorithm for $l_1$ fuzzy c-means and its termination. *Research Report ISE-TR-95-127*, University of Tsukuba.

Morey, L. C., Blashfield, R. K. and Skinner, H. A. (1983) A comparison of cluster analysis techniques within a sequential validation framework. *Multivariate Behavioral Research*, **18**, 309–29.

Morgan, B. J. T. and Ray, A. P. G. (1995) Non-uniqueness and inversions in cluster analysis. *Applied Statistics*, **44**, 117–34.

Moss, W. W. (1967) Some new analytic and graphic approaches to numerical taxonomy, with an example from the Dermanyssidae (Acari). *Systematic Zoology*, **16**, 177–207.

Müller, D. W. and Sawitzki, G. (1991) Excess mass estimates and tests for multimodality. *Journal of the American Statistical Association*, **86**, 738–46.

Mulvey, J. M. and Crowder, H. P. (1979) Cluster analysis: an application of Lagrangian relaxation. *Management Science*, **25**, 329–40.

Murtagh, F. (1983) A survey of recent advances in hierarchical clustering algorithms. *Computer Journal*, **26**, 354–9.

Murtagh, F. (1984) Counting dendrograms: a survey. *Discrete Applied Mathematics*, **7**, 191–9.

Murtagh, F. (1985) A survey of algorithms for contiguity-constrained clustering and related problems. *Computer Journal*, **28**, 82–8.

Murtagh, F. (1995) Interpreting the Kohonen self-organizing feature map using contiguity-constrained clustering. *Pattern Recognition Letters*, **16**, 399–408.

Murtagh, F. (1996) Neural networks for clustering, in *Clustering and Classification* (eds P. Arabie, L. J. Hubert and G. De Soete), World Scientific, Singapore, pp. 235–69.

Murtagh, F. and Hernández-Pajares, M. (1995) The Kohonen self-organizing map method: an assessment. *Journal of Classification*, **12**, 165–90.

Nelson, G. (1979) Cladistic analysis and synthesis: principles and definitions, with a historical note on Adanson's *Famille des Plantes* (1763-1764). *Systematic Zoology*, **28**, 1–21.

Neumann, D. A. (1983) Faithful consensus methods for *n*-trees. *Mathematical Biosciences*, **63**, 271–87.

Ni, L. M. and Jain, A. K. (1985) A VLSI systolic architecture for pattern clustering. *IEEE Transactions on Pattern Analysis and Machine Intelligence*, **PAMI-7**, 80–9.

Oden, N. L. and Shao, K.-T. (1984) An algorithm to equiprobably generate all directed trees with *k* labeled terminal nodes and unlabeled interior nodes. *Bulletin of Mathematical Biology*, **46**, 379–87.

Ohsumi, N. and Nakamura, N. (1989) Space-distorting properties in agglomerative hierarchical clustering algorithms and a simplified method for combinatorial method, in *Data Analysis, Learning Symbolic and Numeric Knowledge* (ed E. Diday), Nova Science Publishers, New York, pp. 103–8.

Olson, C. F. (1995) Parallel algorithms for hierarchical clustering. *Parallel Computing*, **21**, 1313–25.

Pal, N. R. and Bezdek, J. C. (1995) On cluster validity for the fuzzy *c*-means model. *IEEE Transactions on Fuzzy Systems*, **3**, 370–9.

Panayirci, E. and Dubes, R. C. (1983) A test for multidimensional clustering tendency. *Pattern Recognition*, **16**, 433–44.

Papaioannou, T. (1985) Measures of information, in *Encyclopedia of Statistical Sciences, Volume 5* (eds S. Kotz, N. L. Johnson and C. B. Read), Wiley, New York, pp. 391–7.

Papathomas, T. V., Schiavone, J. A. and Julesz, B. (1987) Stereo animation for very large data bases: case study – meteorology. *IEEE Computer Graphics and Applications*, **7**, 18–27.

Payne, R. W. and Preece, D. A. (1980) Identification keys and diagnostic tables: a review (with discussion). *Journal of the Royal Statistical Society*, **A 143**, 253–92.

Pearson, K. (1901) On lines and planes of closest fit to systems of points in space. *Philosophical Magazine (6th Series)*, **2**, 559–72.

Peli, T. and Malah, D. (1982) A study of edge detection algorithms. *Computer Graphics and Image Processing*, **20**, 1–21.

Perruchet, C. (1983) Constrained agglomerative hierarchical classification. *Pattern Recognition*, **16**, 213–7.

Pliner, V. (1996) Metric unidimensional scaling and global optimization. *Journal of Classification*, **13**, 3–18.

Podani, J. (1989) New combinatorial clustering methods. *Vegetatio*, **81**, 61–77.

Pogue, C. A., Rasmussen, E. M. and Willett, P. (1988) Searching and clustering of databases using the ICL distributed array processor. *Parallel Computing*, **8**, 399–407.

Polonik, W. (1995) Measuring mass concentrations and estimating density contour classes – an excess mass approach. *Annals of Statistics*, **23**, 855–81.

Preparata, F. P. and Shamos, M. I. (1988) *Computational Geometry: An Introduction*, Springer, New York.

Quinlan, J. R. (1987) Simplifying decision trees. *International Journal of Man-Machine Studies*, **27**, 221–34.

Quiroz, A. J. (1989) Fast random generation of binary, *t*-ary and other types of trees. *Journal of Classification*, **6**, 223–31.

Ramsay, J. O. (1977) Maximum likelihood estimation in multidimensional scaling. *Psychometrika*, **42**, 241–66.

Ramsay, J. O. (1982) Some statistical approaches to multidimensional scaling data (with discussion). *Journal of the Royal Statistical Society*, A **145**, 285–312.

Rand, W. M. (1971) Objective criteria for the evaluation of clustering methods. *Journal of the American Statistical Association*, **66**, 846–50.

Rao, M. R. (1971) Cluster analysis and mathematical programming. *Journal of the American Statistical Association*, **66**, 622–6.

Rapoport, A. and Fillenbaum, S. (1972) An experimental study of semantic structures, in *Multidimensional Scaling. Theory and Applications in the Behavioral Sciences, Volume II Applications* (eds A. K. Romney, R. N. Shepard and S. B. Nerlove), Seminar Press, New York, pp. 93–131.

Rasmussen, E. M. and Willett, P. (1989) Efficiency of hierarchic agglomerative clustering using the ICL distributed array processor. *Journal of Documentation*, **45**, 1–24.

Régnier, S. (1965) Sur quelques aspects mathématiques des problèmes de classification automatique. *International Computation Centre Bulletin*, **4**, 175–91. Reprinted in *Mathématiques et Sciences Humaines*, **82**, 13–29 (1982).

Reyment, R. A., Blackith, R. E. and Campbell, N. A. (1984) *Multivariate Morphometrics*, 2nd edn, Academic Press, London.

Ripley, B. D. (1996) *Pattern Recognition and Neural Networks*, Cambridge University Press, Cambridge.

Ripley, B. D. and Rasson, J.-P. (1977) Finding the edge of a Poisson forest. *Journal of Applied Probability,* **14,** 483–91.

Robinson, W. S. (1951) A method for chronologically ordering archaeological deposits. *American Antiquity,* **16,** 293–301.

Rohlf, F. J. (1970) Adaptive hierarchical clustering schemes. *Systematic Zoology,* **19,** 58–82.

Rohlf, F. J. (1975a) A new approach to the computation of the Jardine-Sibson overlapping $B_k$ clusters. *Computer Journal,* **18,** 164–8.

Rohlf, F. J. (1975b) Generalization of the gap test for the detection of multivariate outliers. *Biometrics,* **31,** 93–101.

Rohlf, F. J. (1982a) Single-link clustering algorithms, in *Handbook of Statistics, Volume 2: Classification, Pattern Recognition, and Reduction of Dimensionality* (eds P. R. Krishnaiah and L. N. Kanal), North-Holland, Amsterdam, pp. 267–84.

Rohlf, F. J. (1982b) Consensus indices for comparing classifications. *Mathematical Biosciences,* **59,** 131–44.

Rohlf, F. J. (1983) Numbering binary trees with labeled terminal vertices. *Bulletin of Mathematical Biology,* **45,** 33–40.

Rohlf, F. J. and Fisher, D. R. (1968) Tests for hierarchical structure in random data sets. *Systematic Zoology,* **17,** 407–12.

Rosen, D. E. (1978) Vicariant patterns and historical explanation in biogeography. *Systematic Zoology,* **27,** 159–88.

Rosenberg, S. (1982) The method of sorting in multivariate research with applications selected from cognitive psychology and person perception, in *Multivariate Applications in the Social Sciences* (eds N. Hirschberg and L. G. Humphreys), Erlbaum, Hillsdale, NJ, pp. 117–42.

Rosenberg, S. and Kim, M. P. (1975) The method of sorting as a data-gathering procedure in multivariate research. *Multivariate Behavioral Research,* **10,** 489–502.

Roubens, M. (1978) Pattern classification problems and fuzzy sets. *Fuzzy Sets and Systems,* **1,** 239–53.

Rousseeuw, P. J. (1986) A visual display for hierarchical classification, in *Data Analysis and Informatics, IV* (eds E. Diday, Y. Escoufier, L. Lebart, J. Pagès, Y. Schektman and R. Tomassone), North-Holland, Amsterdam, pp. 743–8.

Rousseeuw, P. J. (1987) Silhouettes: a graphical aid to the interpretation and validation of cluster analysis. *Journal of Computational and Applied Mathematics,* **20,** 53–65.

Rowell, A. J., McBride, D. J. and Palmer, A. R. (1973) Quantitative study of Trempealeanian (Latest Cambrian) trilobite distribution in North America. *Bulletin of the Geological Society of America,* **84,** 3429–42.

Rozál, G. P. M. and Hartigan, J. A. (1994) The MAP test for multimodality. *Journal of Classification,* **11,** 5–36.

Rubin, P. A. (1984) Generating random points in a polytope. *Communications in Statistics - Simulation and Computation*, **B 13**, 375-96.

Ruspini, E. H. (1969) A new approach to clustering. *Information and Control*, **15**, 22-32.

Ruspini, E. H. (1970) Numerical methods for fuzzy clustering. *Information Sciences*, **2**, 319-50.

Sammon, J. W. (1969) A non-linear mapping for data structure analysis. *IEEE Transactions on Computers*, **C-18**, 401-9.

Saunders, R. and Funk, G. M. (1977) Poisson limits for a clustering model of Strauss. *Journal of Applied Probability*, **4**, 776-84.

Schoenberg, I. J. (1935) Remarks to Maurice Fréchet's article 'Sur la définition axiomatique d'une classe d'espace distanciés vectoriellement applicable sur l'espace de Hilbert'. *Annals of Mathematics (2nd Series)*, **36**, 724-32.

Schultz, J. V. and Hubert, L. J. (1973) Data analysis and the connectivity of random graphs. *Journal of Mathematical Psychology*, **10**, 421-8.

Scott, A. J. and Symons, M. J. (1971) Clustering methods based on likelihood ratio criteria. *Biometrics*, **27**, 387-97.

Selim, S. Z. and Asultan, K. (1991) A simulated annealing algorithm for the clustering problem. *Pattern Recognition*, **24**, 1003-8.

Shepard, R. N. (1962a) The analysis of proximities: multidimensional scaling with an unknown distance function. I. *Psychometrika*, **27**, 125-40.

Shepard, R. N. (1962b) The analysis of proximities: multidimensional scaling with an unknown distance function. II. *Psychometrika*, **27**, 219-46.

Shepard, R. N. (1963) Analysis of proximities as a technique for the study of information processing in Man. *Human Factors*, **5**, 33-48.

Shepard, R. N. (1972) Psychological representation of speech sounds, in *Human Communication: A Unified View* (eds E. E. David, Jr. and P. B. Denes), McGraw-Hill, New York, pp. 67-113.

Shepard, R. N. (1974) Representation of structure in similarity data: problems and prospects. *Psychometrika*, **39**, 373-421.

Shepard, R. N. and Arabie, P. (1979) Additive clustering: representation of similarities as combinations of discrete overlapping properties. *Psychological Review*, **86**, 87-123.

Sibson, R. (1972) Multidimensional scaling in theory and practice, in *Les Méthodes Mathématiques de l'Archéologie*, Centre d'Analyse Documentaire pour l'Archéologie, Marseille, pp. 43-73.

Sibson, R. (1973) SLINK: an optimally efficient algorithm for the single-link cluster method. *Computer Journal*, **16**, 30-4.

Sibson, R. (1979) Studies in the robustness of multidimensional scaling: perturbational analysis of classical scaling. *Journal of the Royal Statistical Society*, **B 34**, 311-49.

Sibson, R., Bowyer, A. and Osmond, C. (1981) Studies in the robustness of multidimensional scaling: Euclidean models and simulation studies. *Journal of Statistical Computation and Simulation*, **13**, 273–96.

Smith, S. P. and Jain, A. K. (1984) Testing for uniformity in multidimensional data. *IEEE Transactions on Pattern Analysis and Machine Intelligence*, **PAMI-6**, 73–81.

Sneath, P. H. A. (1957) The application of computers to taxonomy. *Journal of General Microbiology*, **17**, 201–26.

Sneath, P. H. A. (1966) A comparison of different clustering methods as applied to randomly-spaced points. *Classification Society Bulletin*, **1(2)**, 2–18.

Sneath, P. H. A. (1967) Some statistical problems in numerical taxonomy. *The Statistician*, **17**, 1–12.

Sneath, P. H. A. (1969) Evaluation of clustering methods (with discussion), in *Numerical Taxonomy* (ed A. J. Cole), Academic Press, London, pp. 257–71.

Sneath, P. H. A. and Sokal, R. R. (1973) *Numerical Taxonomy*, Freeman, San Francisco.

Sokal, R. R. and Michener, C. D. (1958) A statistical method for evaluating systematic relationships. *University of Kansas Science Bulletin*, **38**, 1409–38.

Sokal, R. R. and Rohlf, F. J. (1962) The comparison of dendrograms by objective methods. *Taxon*, **11**, 33–40.

Sokal, R. R. and Rohlf, F. J. (1980) An experiment in taxonomic judgment. *Systematic Botany*, **5**, 341–65.

Sokal, R. R. and Rohlf, F. J. (1981) Taxonomic congruence in the Leptodomorpha re-examined. *Systematic Zoology*, **30**, 309–25.

Sparks, R., Adolphson, A. and Phatak, A. (1997) Multivariate process monitoring using the dynamic biplot. *International Statistical Review*, **65**, 325–49.

Spence, I. and Domoney, D. W. (1974) Single subject incomplete designs for nonmetric multidimensional scaling. *Psychometrika*, **39**, 469–90.

Spence, I. and Lewandowsky, S. (1989) Robust multidimensional scaling. *Psychometrika*, **54**, 501–13.

Spence, N. A. (1968) A multifactor uniform regionalization of British counties on the basis of unemployment data for 1961. *Regional Studies*, **2**, 87–104.

Sriram, N. (1990) Clique optimization: a method to construct parsimonious ultrametric trees from similarity data. *Journal of Classification*, **7**, 33–52.

Steel, M. (1992) The complexity of reconstructing trees from qualitative characters and subtrees. *Journal of Classification*, **9**, 91–116.

Steel, M. and Warnow, T. (1993) Kaikoura tree theorems: computing the maximum agreement subtree. *Information Processing Letters*, **48**,

77–82.

Stéphan, V. (1996) Description de classes par des assertions, in *Analyse des Données Symboliques, Tome 1*, École d'Été, Université Paris - IX Dauphine, pp. 241–53.

Stinebrickner, R. (1984) *s*-Consensus trees and indices. *Bulletin of Mathematical Biology*, **46**, 923–35.

Strauss, D. J. (1975) A model for clustering. *Biometrika*, **62**, 467–75.

Sun, L.-X., Xie, Y.-L., Song, X.-H., Wang, J.-H. and Yu, R.-Q. (1994) Cluster analysis by simulated annealing. *Computers and Chemistry*, **18**, 103–8.

Sutcliffe, J. P. (1994) On the logical necessity and priority of a monothetic conception of class, and on the consequent inadequacy of polythetic accounts of category and categorization, in *New Approaches in Classification and Data Analysis* (eds E. Diday, Y. Lechevallier, M. Schader, P. Bertrand and B. Burtschy), Springer, Berlin, pp. 55–63.

Symons, M. J. (1981) Clustering criteria and multivariate normal mixtures. *Biometrics*, **37**, 35–43.

Tong, T. T. H. and Ho, T. B. (1991) A method for generating rules from examples and its application, in *Symbolic-Numeric Data Analysis and Learning* (eds E. Diday and Y. Lechevallier), Nova Science, New York, pp. 493–504.

Torgerson, W. S. (1952) Multidimensional scaling: I. Theory and method. *Psychometrika*, **17**, 401–19.

Toussaint, G. T. (1980a) Pattern recognition and geometrical complexity. *Proceedings of the 5th International Conference on Pattern Recognition*, 1324–47.

Toussaint, G. T. (1980b) The relative neighbourhood graph of a finite planar set. *Pattern Recognition*, **12**, 261–8.

Trauwaert, E. (1987) $L_1$ in fuzzy clustering, in *Statistical Data Analysis Based on the $L_1$-Norm and Related Methods* (ed Y. Dodge), North-Holland, Amsterdam, pp. 417–26.

Trauwaert, E., Kaufman, L. and Rousseeuw, P. (1991) Fuzzy clustering algorithms based on the maximum likelihood principle. *Fuzzy Sets and Systems*, **42**, 213–27.

Tsai, H.-R., Horng, S.-J., Lee, S.-S., Tsai, S.-S. and Kao, T.-W. (1997) Parallel hierarchical clustering algorithms on processor arrays with a reconfigurable bus system. *Pattern Recognition*, **30**, 801–15.

Tucker, L. R. (1964) The extension of factor analysis to three-dimensional matrices, in *Contributions to Mathematical Psychology* (eds N. Frederiksen and H. Gulliksen), Holt, Rinehart and Winston, New York, pp. 109–27.

Tukey, P. A. and Tukey, J. W. (1981a) Preparation; prechosen sequence of views, in *Interpreting Multivariate Data* (ed V. Barnett), Wiley, Chichester, pp. 189–213.

Tukey, P. A. and Tukey, J. W. (1981b) Summarization; smoothing; supplemented views, in *Interpreting Multivariate Data* (ed V. Barnett), Wiley, Chichester, pp. 245–75.

Tversky, A. (1977) Features of similarity. *Psychological Review*, **84**, 327–52.

Van Cutsem, B. and Ycart, B. (1998) Indexed dendrograms on random dissimilarities. *Journal of Classification*, **15**, 93–127.

Van Ness, J. W. (1973) Admissible clustering procedures. *Biometrika*, **60**, 422–4.

van Rijsbergen, C. J. (1970) A clustering algorithm. *Computer Journal*, **13**, 113–5.

Vinod, H. D. (1969) Integer programming and the theory of grouping. *Journal of the American Statistical Association*, **64**, 506–19.

Ward, J. H. , Jr. (1963) Hierarchical grouping to optimize an objective function. *Journal of the American Statistical Association*, **58**, 236–44.

Watson, L., Williams, W. T. and Lance, G. N. (1966) Angiosperm taxonomy: a comparative study of some novel numerical techniques. *Botanical Journal of the Linnean Society*, **59**, 491–501.

Webster, R. and Burrough, P. A. (1972) Computer-based soil mapping of small areas from sample data II. Classification smoothing. *Journal of Soil Science*, **23**, 222–34.

Welch, W. J. (1982) Algorithmic complexity: three NP-hard problems in computational statistics. *Journal of Statistical Computation and Simulation*, **15**, 17–25.

Whaley, R. and Hodes, L. (1991) Clustering a large number of compounds. 2. Using the Connection machine. *Journal of Chemical Information and Computer Science*, **31**, 345–7.

Wilkinson, M. (1994) Common cladistic information and its consensus representation: reduced Adams and reduced cladistic consensus trees and profiles. *Systematic Biology*, **43**, 343–68.

Wilkinson, M. (1995) More on reduced consensus methods. *Systematic Biology*, **44**, 435–9.

Williams, W. T. (Ed) (1976) *Pattern Analysis in Agricultural Science*, CSIRO, Melbourne / Elsevier, Amsterdam.

Williams, W. T. and Lambert, J. M. (1959) Multivariate methods in plant ecology I. Association analysis in plant communities. *Journal of Ecology*, **47**, 83–101.

Williamson, M. H. (1978) The ordination of incidence data. *Journal of Ecology*, **66**, 911–20.

Windham, M. P. (1982) Cluster validity for the fuzzy c-means clustering algorithm. *IEEE Transactions on Pattern Analysis and Machine Intelligence*, **PAMI-4**, 357–63.

Windham, M. P. (1987) Parameter modification for clustering criteria. *Journal of Classification*, **4**, 191–214.

Wirth, M., Estabrook, G. F. and Rogers, D. J. (1966) A graph theory model for systematic biology, with an example for the Oncidiinae (Orchidaceae). *Systematic Zoology*, **15**, 59–69.

Wishart, D. (1969a) Mode analysis: a generalization of nearest neighbour which reduces chaining effects (with discussion), in *Numerical Taxonomy* (ed A. J. Cole), Academic Press, London, pp. 282–311.

Wishart, D. (1969b) An algorithm for hierarchical classifications. *Biometrics*, **25**, 165–70.

Wishart, D. (1987) *Clustan User Manual*, 4th edn, Computing Laboratory, University of St Andrews.

Wright, W. E. (1973) A formalization of cluster analysis. *Pattern Recognition*, **5**, 273–82.

Xie, X. L. and Beni, G. (1991) A validity measure for fuzzy clustering. *IEEE Transactions on Pattern Analysis and Machine Intelligence*, **13**, 841–7.

Young, F. W., Faldowski, R. A. and McFarlane, M. M. (1993) Multivariate statistical visualization, in *Handbook of Statistics, Volume 9 Computational Statistics* (ed C. R. Rao), North-Holland, Amsterdam, pp. 959–98.

Young, G. and Householder, A. S. (1938) Discussion of a set of points in terms of mutual distances. *Psychometrika*, **3**, 19–22.

Zahn, C. T. (1971) Graph-theoretical methods for detecting and describing gestalt clusters. *IEEE Transactions on Computers*, **C-20**, 68–86.

Zeng, G. and Dubes, R. C. (1985) A test for spatial randomness based on $k$-NN distances. *Pattern Recognition Letters*, **3**, 85–91.

# Author index

# Subject index

Absence of cluster structure, tests for 188–90
Adaptive clustering procedures 48, 56–8
Additive clustering 122–3
Adequacy of graphical representation 148, 154, 156, 160, 175
Admissibility 98–100, 188
Agglomerative algorithms 60, 63–4, 78–90, 117–20, 127, 141–2
Aims of classifications 5–6
Algorithms
  agglomerative 60, 63–4, 78–90, 117–20, 127, 141–2
  branch-and-bound 40, 52, 76
  direct optimization 75–7, 105–6, 120–2, 127, 154, 160, 166
  divisive 90–1, 120, 130–4, 208–9
  dynamic programming 40, 76, 119, 165
  general agglomerative 78–80, 83, 87–9, 118–9
  hybrid 44, 54–7
  incremental 90, 135, 201
  iterative relocation 41–9, 56, 119, 130, 144, 170
  methods and 96
  single pass 42–3
Angular separation 21, 29
Assignment 3, 5, 55–8, 67, 133–4, 172, 207–9
Asymmetry 15
Atypical objects 43, 204–8

Axiomatic approach 87, 97–8

$B_k$ clusters 123–5
Ball clusters 58–60, 194
Banners 73
Bayesian assumptions 67–8
Bhattacharyya's coefficient 22–3
Binary trees 70, 72, 201
Binary variables 17–8, 20, 29, 128–34
Biplots 172–82
Branch-and-bound algorithms 40, 52, 76

$c$-means clustering, see Iterative relocation algorithms
Canberra metric 20, 30
Centroid clustering 79, 87, 89
Choice of clustering strategy 96–100
City block metric 20, 30, 159
Classical scaling, see Principal coordinates analysis
Classification(s)
  admissibility approach to 98–100, 188
  aims of 5–6
  axiomatic approach to 87, 97–8
  constrained 93, 115–21, 146, 170
  desiderata of 3–4
  distortion of 97–8, 200
  history of 5
  maximal predictive 128–30
  overlapping 3, 60, 121–7